Environmental Footprints and Eco-design of Products and Processes

Series editor

Subramanian Senthilkannan Muthu, SGS Hong Kong Limited, Hong Kong, Hong Kong

This series aims to broadly cover all the aspects related to environmental assessment of products, development of environmental and ecological indicators and eco-design of various products and processes. Below are the areas fall under the aims and scope of this series, but not limited to: Environmental Life Cycle Assessment; Social Life Cycle Assessment; Organizational and Product Carbon Footprints; Ecological, Energy and Water Footprints; Life cycle costing; Environmental and sustainable indicators; Environmental impact assessment methods and tools; Eco-design (sustainable design) aspects and tools; Biodegradation studies; Recycling; Solid waste management; Environmental and social audits; Green Purchasing and tools; Product environmental footprints; Environmental management standards and regulations; Eco-labels; Green Claims and green washing; Assessment of sustainability aspects.

More information about this series at http://www.springer.com/series/13340

Miguel Angel Gardetti
Subramanian Senthilkannan Muthu
Editors

Sustainable Luxury

Cases on Circular Economy
and Entrepreneurship

 Springer

Editors
Miguel Angel Gardetti
Center for Studies on Sustainable Luxury
Buenos Aires
Argentina

Subramanian Senthilkannan Muthu
SGS Hong Kong Limited
Hong Kong
Hong Kong

ISSN 2345-7651   ISSN 2345-766X (electronic)
Environmental Footprints and Eco-design of Products and Processes
ISBN 978-981-13-4464-0   ISBN 978-981-13-0623-5   (eBook)
https://doi.org/10.1007/978-981-13-0623-5

Printed on acid-free paper

This Springer imprint is published by the registered company Springer Nature Singapore Pte Ltd. part of Springer Nature
The registered company address is: 152 Beach Road, #21-01/04 Gateway East, Singapore 189721, Singapore

Preface

Luxury depends on cultural, economic or regional contexts which transform luxury into an ambiguous concept. Also, according to Ricca and Robins (2012), luxury is a source of inspiration, controversy, admiration and considerable economic success. But luxury is becoming more about helping people to express their deepest values. So, sustainable luxury would not only be a vehicle for more respect for the environment and social development, but it will also be a synonym of culture, art and innovation of different nationalities, maintaining the legacy of local craftsmanship (Gardetti 2011).

"The circular economy refers to an industrial economy that is restorative by intention; aims to rely on renewable energy; minimises, tracks, and eliminates the use of toxic chemicals; and eradicates waste through careful design. The term goes beyond the mechanics of production and consumption of goods and services in the areas that it seeks to redefine" (Ellen Mac Arthur Foundation 2013, p. 22). Examples include rebuilding capital, including social and natural, and the shift from consumer to user (Ellen Mac Arthur Foundation 2013, p. 22).

On the other hand, there are people with a profound perspective towards environmental and social issues and who are well motivated to "break" the rules and promote disruptive solutions to these issues, and most of them are entrepreneurs. These individuals have a number of different roles to play in entrepreneurship or intrapreneurship and innovation, from the imaginative act of setting up a new venture. This involves cognitive and motivational characteristics.

This book shows cases of circular economy related to entrepreneurs, always within the framework of sustainable luxury, as shown below.

The Book

The book begins with a paper by Kalina Yingnan Deng "Vestire: Social Divesting and Impact Investing in New Materialism". Taking a new materialist approach to people's lives and the values of clothes, the author shows how women are oriented towards a particular type of (in)vestment that is under-recognised and undervalued under the lights cast by the pursuit of fashion as superficial. In other words, at the root of the term *investment* is *vestire*, to clothe, already suggesting that popular (mis)conceptions about fashion's frivolity lack *material* evidence. Using auto-ethnography as the author's primary method and the Buffalo Exchange second-hand chain of stores and particularly the East Village, Manhattan location as the author primary field site, she illustrates the cycling through of clothing and the changes in these commodities' value. As I show in this case study, in the second-hand market for luxe designer garments, the concept of value changes as quickly as it moves through hands in the handed-down, preowned, preloved neb-ulas that make *value* itself amorphous. Moreover, as shown by the entrepreneurial example of Brass, alternative extra-industry solutions to accessing such (in)vest-ments can close the wardrobe gap for women.

Moving on, Carmela Donato, Cesare Amatulli and Matteo De Angelis developed "Responsible Luxury Development: A Study on Luxury Companies' CSR, Circular Economy, and Entrepreneurship". In this chapter the authors discuss how luxury brands can build their success on corporate social responsibility (CSR), leveraging specifically on the paradigm of circular economy. The idea elaborated on the chapter is that luxury and sustainability are not conflicting concepts, as many believe, but they are positively correlated, in as much as the quintessential char-acteristics of luxury goods make them potentially more sustainable than mass-market goods. Through the discussion of four case studies of luxury brands operating in the sectors of fashion (Brunello Cucinelli, Gucci and Stella McCartney brands) and food (Godiva), the authors point out that the reuse of tangible resources, such as money generated by business activities and raw material, can be a very solid basis for building market success, as well as for broadening the positive contribution luxury brands can make to the environment, the employees, the local community of producers, and, as a consequence, to the society at large. A common feature of all the cases discussed is represented by the key role played by the entrepreneur (often the company's founder) in fostering a balance between brand prestige and sustainability.

The chapter called "Challenging Current Fashion Business Models: Entrepreneurship Through Access-Based Consumption in the Second-Hand Luxury Garment Sector Within a Circular Economy" was written Shuang Hu, Claudia E. Henninger, Rosy Boardman and Daniella Ryding. The purpose of this chapter is to research about the drivers of (non)participation in access-based con-sumption and the underpinning motives of becoming (or not) a micro-entrepreneur within the circular economy. Peer-to-peer platforms and drivers of (non)participa-tion within the context of the UK's second-hand luxury market are currently under-researched. This chapter is exploratory in nature and utilises a qualitative research approach.

In turn, Sheetal Jain and Sita Mishra, in their paper "Sadhu—On the Pathway of Luxury Sustainable Circular Value Model" research into the company "Natweave Textile Studio". This is a textile company founded by Indian textile designer Subhabrata Sadhu in 2009, with a yearning to conserve the rich heritage of the rarest and finest cashmere by using the traditional skills of native Kashmiri artisans. The company specialises in producing high-end and exclusive Pashmina scarves and shawls with a focus on entirely pure, handmade and natural production process. Sadhu sources the finest Pashmina fibres from Pashmina goats reared in its natural habitat in Changthang plateau in the Kashmir region. This study aims to develop a luxury sustainable circular value (LSCV) model that integrates the values of four stakeholders—entrepreneur, organisation, customers and society. LSCV model is used as a tool to examine how "Natweave Textile Studio" contributes to creating sustainable circular value and thus adding to the sustainable development of the company and society.

Following, Ansgar Igelbrink, Albin Kälin, Marko Krajner and Roman Kunič's paper titled "Cradle to Cradle®—Parquet for Generations: Respect Natural Resources and Offers Preservation for the Future" focuses on wood. Today, the use of wood in architecture is becoming fashionable. The authors feel strongly that developments in wood products and timber construction will shape the future of sustainable development. Wood is still one of the most accessible materials; we can smell, hear, touch, and see (it is pleasing to the eye) natural wood. A similar development can be seen in the façades of modern structures, where imitation wood is increasingly being used. Sustainable luxury products incorporate extraordinary aesthetics, handle, care, function and, in addition to being sustainable, they need to be safe for humans, society and the environment. Resources, especially natural resources, are scare and need to be protected in changing the design of products' uses according to "Cradle to Cradle®: Remaking the way we make things" and towards a circular economy. For companies, this implies entrepreneurship to tackle the large impact on the change of behaviour, culture, marketing and business models in closing the loop and taking the goods back from the user. The case study, Cradle to Cradle®—Parquet for Generations (of Bauwerk Parkett): Respect Resources and Preservation for the Future, illustrates a successful lighthouse example from the industry.

In the next chapter, "Trends of Sustainable Development Among Luxury Industry" by Jitong Li and Karen K. Leonas, the authors clarify that under increasing pressure to implement sustainable development throughout the industry, some new luxury entrepreneurs are emerging with remarkable perspectives on sustainable development. They break the traditional business innovation known to the luxury sector and are implementing the concept of sustainable development as a direction in their business strategies. In addition, they are moving towards developing a circular economy to realise "sustainability" in their supply chains. This chapter discusses the redefinition of luxury, trends in the luxury market, adoption of sustainability among luxury brands and consumers, disruptive business, model innovation and the circular economy. Finally, a case study on sustainable luxury swimwear entrepreneurs is presented.

Completing the book, Sabine Weber prepared a paper titled "A Circular Economy Approach in the Luxury Fashion Industry: A Case Study of Eileen Fisher". This study used a case study approach to observe and analyse the circular economy business model of Eileen Fisher (EF), New York. This study explores how the company has developed its take-back programme and how this programme led to the development of recycling operations at EF. In 2017, twelve semi-structured interviews were conducted with employees from EF, representing different departments and operating at various functions in the company. Their responses were analysed according to a content analysis method to outline EF's approaches to both luxury fashion and circular economy, and additional data from the company were collected. The results were summarised in the strengths, weaknesses, opportunities and threats (SWOT) analysis to show the advantages and challenges a company faces when introducing the circular economy concept.

It is important to highlight that all of these diverse contributions represent a great step forward in expanding the insights in the field of sustainable management of luxury. It is certainly the most comprehensive collection of writings on these subjects to date. Note that this initiative has received a large international response, and it is expected to continue to stimulate further debate.

Buenos Aires, Argentina   Miguel Angel Gardetti
Hong Kong   Subramanian Senthilkannan Muthu

Bibliography

Ellen MacArthur Foundation. (2013). *Towards the Circular Economy—Economic and business rationale for an accelerated transition*. Ellen MacArthur Foundation, Cowes.
Gardetti, M. A. (2011). Sustainable luxury in Latin America. In *Conference delivered at the Seminar Sustainable Luxury & Design within the framework of IE*—Instituto de Empresa—Business School MBA, Madrid, Spain.
Ricca, M., & Robins, R. (2012). *Meta-luxury—Brands and the culture of excellence*. New York: Palgrave Macmillan.

Contents

Vestire: Social Divesting and Impact Investing in New Materialism

Kalina Yingnan Deng

Abstract Fashion, with its social gloss of being the most transient of commodities in postmodern Western capitalistic societies, undergirds the hunt for the next "it" item in our modern liquid worlds (Bauman 2006). In fashion, as post-Marxist socialist Zygmunt Bauman argues, there is a perpetuum mobile, or social dynamic, in which progress is articulated as each individual's avoidance of exclusion (Marx 2015 [1867]: 85). As individuals fear exclusion, the individual's relation to fashion fulfills a continuous cycle of becoming, a hedonic treadmill with temporary happiness, via materialistic consumption, as its constant goal (Bauman 2010). However, the rise of secondhand markets in the liquid West and consumers' engagement in both the brick-and-mortar and ecommerce varieties thereof undercut dialectical revolutions of consumption habits for the newest and shiniest of baubles. Taking a new materialist approach to lives and values of clothes, I show how females are oriented toward a particular type of (in)vestment that is underrecognized and undervalued under the lights that cast the pursuit of fashion as superficial. In other words, at the root of the term *investment* is *vestire*, to clothe, already suggesting that popular (mis)conceptions about fashion's frivolity lacks *material* evidence. Using auto-ethnography as my primary method and the Buffalo Exchange secondhand chain of stores and particularly the East Village, Manhattan location as my primary field site, I illustrate the cycling through of clothing and the changes in these commodities' value. I complicate the easy narratives that Western political economists have used to explain away Marx's theory on value and commodity fetishism. The lifecycle of commodities, as symptomized in the classic macroeconomic example of guns versus butter, shows a production possibility frontier circumscribed by the assumption that consumption is a one-time deal. Yet, consuming and wearing secondhand clothing can be upheld by new materialism as a socially impactful (in)vestment. As I show in this case study, in the secondhand market for luxe designer garments, the concept of value changes

K. Y. Deng (✉)
Parsons School of Design, New York City, NY, USA
e-mail: kydeng@newschool.edu

© Springer Nature Singapore Pte Ltd. 2019
M. A. Gardetti and S. S. Muthu (eds.), *Sustainable Luxury*,
Environmental Footprints and Eco-design of Products and Processes,
https://doi.org/10.1007/978-981-13-0623-5_1

1

quickly as it moves through hands in the handed-down, preowned, preloved nebulas that make *value* itself amorphous. Moreover, as shown by the entrepreneurial example of Brass, alternative extra-industry solutions to accessing such (in)vestments can close the wardrobe gap for women.

Keywords Secondhand markets · Buffalo exchange · Commodity fetishism Luxury fashion · Wardrobe gap · Value · Circular economy · Sustainability

1 Introduction

I am a materialist. Beyond that, I want to reclaim the word "materialist" and its derivative, "materialism." To be clear, being a "materialist" differs from being "materialistic." As a materialist, I value the physical materials and materiality of things. Contra those who are materialistic, I value less any name brands and logo-bedecked garb. As a materialist, I invest in my wardrobe and in myself. Taking a new materialist approach to lives and values of clothes, I show how females are oriented toward a particular type of (in)vestment that is underrecognized and undervalued under the lights that cast the pursuit of fashion as superficial. In other words, at the root of the term *investment* is *vestire*, to clothe, already suggesting that popular (mis)conceptions about fashion's frivolity lacks *material* evidence.

To wit, fashion, with its social gloss of being the most transient of commodities in postmodern Western capitalistic societies, undergirds the hunt for the next "it" item in our modern liquid worlds (Bauman 2006). In fashion, as post-Marxist socialist Zygmunt Bauman argues, there is a perpetuum mobile, or social dynamic, in which progress is articulated as each individual's avoidance of exclusion. As individuals fear exclusion, the individual's relation to fashion fulfills a continuous cycle of becoming, a hedonic treadmill with temporary happiness, via materialistic consumption, as its constant goal (Bauman 2010). However, the rise of secondhand markets in the liquid West and consumers' engagement in both the brick-and-mortar and ecommerce varieties thereof undercut dialectical revolutions of consumption habits for the newest and shiniest of baubles.

Using auto-ethnography as my primary method and the Buffalo Exchange secondhand chain of stores and particularly the East Village, Manhattan location as my primary field site,[1] I illustrate the cycling through of clothing and the changes in these commodities' value. I complicate the easy narratives that Western political economists have used to explain away Marx's theory on value and commodity

---

[1] I have regularly sold and shopped at Buffalo Exchange since late 2010, starting with the Boston, Massachusetts area locations. Between 2010 and 2016, I suspect that I have spent roughly equal times at both the Coolidge Corner, Allston location as well as the Davis Square, Somerville location. Since moving to New York City in August 2016, I have visited and purchased items from four of the five Buffalo Exchange locations (East Village and Chelsea in Manhattan as well as Williamsburg and Boerum Hill, Brooklyn locations but not the Astoria, Queens location) in the New York City area. I have sold my wares at the East Village, Williamsburg, and Boerum Hill locations.

fetishism. The lifecycle of commodities, as symptomized in the classic macroeconomic example of guns versus butter, shows a production possibility frontier circumscribed by the assumption that consumption is a one-time deal. Yet, consuming and wearing secondhand clothing can be upheld by new materialism as a socially impactful (in)vestment. As I show in this case study, in the secondhand market for luxe designer garments, the concept of value changes quickly as it moves through hands in the handed-down, preowned, preloved nebulas that make *value* itself amorphous. Moreover, as shown by the entrepreneurial example of Brass, alternative extra-industry solutions to accessing such (in)vestments can close the wardrobe gap for women.

This chapter is organized as follows: I first provide an overview of Marx's theory on value. I then reconsider how secondhand markets for luxury fashion, as shown by my auto-ethnography of the Buffalo Exchange chain, complicate Marx's critique of value. I then extend my auto-ethnography into an in-depth analysis of clothing as (in)vestments in the fashioned self. Finally, I offer the entrepreneurial example of Brass, which seeks to address the wardrobe investment gap for women. I close with reflections on a new materialist beginning for fashion.

2 Marx on Value

In the first two chapters of the first volume of *Capital*, Marx outlines the terms and definitions that underpin his version of the labor theory of value.[2] He starts with the definition of the commodity. In Marx's view, a commodity is an object that is useful and external to our person and that can be exchanged on the market. This presumes that there is a market on which commodities can be exchanged. This also presumes that there exists a social division of labor, where each person produces different commodities for exchange with one another on the market. Commodities contain two types of values: use-value and exchange value. Use-value is linearly understood as the utility of the object (Marx 2015 [1867]: 27). Exchange value is reductively understood as price but entails a deeper understanding of the relative value of one commodity versus another. For Marx, what determines the exchange value of a commodity is the labor input in the production of the commodity. Here, labor is constrained as the socially necessary labor needed to produce the commodity "under the normal conditions of production, and with the average degree of skill and intensity prevalent at the time" (Marx 2015 [1867]: 29). In this view, idle or less-than-average skilled hours spent on production should not artificially increase the value of a commodity. In sum, a commodity is something useful produced for the sake of exchange in the market at the exchange value derived from the socially necessary labor time needed to produce it.

---

[2]Given the space allotted in this paper, it would not be possible to provide a full account of Marxian economics. Rather, I endeavor to provide an overview of Marx's view on the commodity and commodity fetishism, as it is relevant to the discussion and analysis provided in this paper.

At this junction, Marx makes a two-step leap in order to claim how the exchange values of different commodities are determined. The steps are as follows:

1. A quantity of a commodity A must be equal to a quantity of another commodity B.

$$A \quad = B$$
$$1 \text{ coat} = 20 \text{ yards of linen}$$

2. To disrupt the vicious cycle of two commodities' interchange against each other due to inflation in exchange value of a given commodity, there must be a third commodity that can be exchanged for either of the other two commodities.

$$A \quad = B \quad\quad\quad\quad = C$$
$$1 \text{ coat} = 20 \text{ yards of linen} = 40 \text{ pounds of coffee}$$

This means that some quantity of coat can equal some quantity of linen, coffee, corn, or tea. Ergo, his general form for value is created (Marx 2015 [1867]: 44). As these commodities can be exchanged at a particular relative rate, all these commodities—one coat, twenty yards of linen, and forty pounds of coffee—can be exchanged, or now understood as bought, for some amount of money, e.g., two ounces of gold (Marx 2015 [1867]: 47). Now, we have Marx's money form for value.

Marx then introduces his critique of the fetishism of commodities in capitalist society. For him, the value of a commodity resides in the immaterial social relations between objects that show its objectified form in the process of exchange of one commodity with another. The immaterial social relationship refers to the socially necessary labor time that was inputted into the process of making the object (Marx 2015 [1867]: 48, 52). Therefore, Marx's critique of commodity fetishism is inextricably tied to his views that workers have been alienated from their labor, and the commodity conceals the social relationship between the producer and the capitalist. In reductionist terms, for Marx, commodities are overvalued for immaterial values undergirded by capitalistic motivations (e.g., modern conceptions of "brand value") and yet devalued for its material values (e.g., labor production).

3   (Re)Selling Marx on Value

From graduating from college in May 2014 through the end of the year, I worked a second job in retail to pay off my undergraduate loans. During the seven months that I worked at the flagship store of a contemporary luxury retailer on Boston's ritzy Newbury Street, I learned how pricing models in the retail game fuel the capitalist superstructure that Marx critiqued. At this retailer, the wholesale price was half of the manufacturer's suggested retail price (MSRP), and our employee's clothing allowance from the company was calculated based on the wholesale price of the clothing items. Using our seasonal clothing allowance, we could "buy" a couple of the season's lower-priced items for our work uniform. As employees, we also received a hefty additional 40% off all items, even on sale items. On several occasions, I purchased final sale items at 60% off, with my additional employee discount, for a grand total of 76% off of the MSRP. During the holidays, there were employee-only sales on warehouse items from previous seasons. Through those sales, I was able to get several fur-trimmed coats at 85% off MSRP.

From my experience at this retailer, I learned how inflated MSRPs must be. Though I was not privy to the exact costs of production for a given garment, I believe that it is logical to assume that even by offering its employees additional discounts on already discounted items, the company is earning net profits on sales to its customer base/brand ambassadors. Moreover, my "fieldwork" at this retailer and my interactions with my colleagues who hailed from other purveyors of luxury fashions, e.g., Barneys, Bally, Armani, Gucci, Valentino, and Aquascutum, elucidated that such high-end brands wanted to brand their employees with "the look" of the company. At this retailer, it meant minimal makeup, sleeked hair, subdued patterns and tones, and no flashy logos. Aesthetic value, the currency for looking cohesive with the brand image, is what truly sold at this retailer—that a simple black cashmere wool blend cardigan is worth $400.

Here, Marx's analysis of the money-commodity-money (M-C-M) circuit and the corresponding commodity-money-commodity (C-M-C) circuit becomes relevant. For Marx, the M-C-M circulation which begins and ends in money is oriented toward exchange value (Marx 2015 [1867]: 106). As also seen from my experience with traditional fashion retail, the model of value exemplifies how Marx tailors the M-C-M circulation into M-C-M′ where M′ is equal to M plus the DM surplus value afforded by the brand and aesthetic values collapsed together in liquid capitalism (Marx 2015 [1867]: 106). At this retailer, the equation of value circulation well fits Marx's critique of commodity fetishism, as such

$$M'(\text{MSRP}) = M(\text{warehouse price}) + DM(\text{surplus value} = \text{brand/aesthetic value})$$

*where the warehouse price already represents the M′ of a previous equation in which M would be the actual cost of materials and labor.

In the circulation of M becoming M′, the magicalities of capitalism comes to play in metamorphosing simple exchange to involve such complex machinations of surplus value at various stages with the Fashion (with a capital F) production of commodities. As made explicit by Marx, "M-C-M′ is therefore in reality the general formula of capital as it appears *prima facie* within the sphere of circulation (Marx 2015 [1867]: 108).

In his piece *"NEW" Collection* (2017) for the *fashion after Fashion* exhibit at the Museum of Arts and Design in New York City, Ryohei Kawanishi complicates easy narratives around value by not only making elements of the usually hidden design process visible but also by retagging secondhand garments originally created

Fig. 1 Louis Vuitton x Rei Kawakubo $2,790 Sak Plat with very large holes in it. Photograph from Louis Vuitton, shot by Jennifer Livingston, "who uses light and shadow to emphasize the new form of the bag that has been created and the way that the traditions of design have been intentionally broken."

by the likes of Martin Margiela and Helmut Lang with a R.K. label.[3] As explained by exhibit curator Hazel Clark, the R.K. label not only riffs off of Kawanishi's own name but also plays off of Rei Kawakubo's collaboration with Louis Vuitton, in which she "adds value" to the house's signature monogrammed Sak Plat tote by putting very large holes in it and literally decreasing its material value. Yet, the Rei Kawakubo collaboration Sak Plat with a R.K. monogrammed inner sack sells for nearly twice as much as the regular monogram Sak Plat (roughly $1,400 to $1,700 across global markets). By playing off of the (non-Rei Kawakubo) R.K. label value addition, Kawanishi shows how the clothes are now devalued, downgraded from Maison Margiela to nameless abasement (Fig. 1).

In the secondhand clothing market, Marx's theory on the production and value of commodities for exchange in the market can no longer remain linear but becomes labyrinthine. Perhaps, to double riff off of Kawanishi and Kawakubo × Vuitton, the understood values of commodities no longer follow the simple articulation of value (money) being artificially inflated by capitalistic, materialistic practices. Against capitalistic drive, secondhand markets and the connotation of purchasing worn items have a punk element, a subcultural tone. Contra both ritzy luxury consignment operations and charity thrift stores, the Buffalo Exchange's gestalt and philosophy are grounded in the counter-culture revolution of the 1970s, as instilled by its founders in their first Tucson, Arizona shop in 1974. From my experiences shopping in present day Buffalo Exchange locations in both New York City and Boston, it often feels as if a black clothing base, hair dyed in carnival colors, tattoo sleeves, and ringed noses

---

[3]To wit, Kawanishi does not design under a R.K. label but rather under the label Landlord.

were pre-requisites for employment at the retail chain.[4] Unlike the bleach-toothed guardians of the Armani's, Valentino's, and Gucci's of the Fashion System, Buffalo Exchange employees are gritty and "real" people who do not typically brunch at Stephanie's on Newbury Street on their Saturdays off. Their taste is not subjective to one brand gestalt but to the demands of the micro-economy of their particular Buffalo Exchange location. The kinds of clothes that may move through one store may not be the same as the items that fly off the rack at another location. As shown on the chain's website, the ecosystem of the neighborhood characterizes each store. For the East Village store near Parsons, the location is described on the official website as such

> Stretching from Third Avenue to the East River in Manhattan, the East Village has contin-uously been an artistic hub in New York City—from the early theater district days, to the migration of the Beatniks in the 50s, to Andy Warhol's infamous art films in the 60s and the punk movement in the 70s. Today, the store attracts the true inhabitants of this unique New York neighborhood; artists, models, stylists, and musicians all come to sell and shop, showcasing an array of unique and original pieces (2017).

Therefore, in the secondhand market of Buffalo Exchange, the target consumer base is sprawling rather than concentrated as with retailers in the traditional Fashion System. Aesthetic value is devalued in that it is harder to pinpoint the exact taste of the wide swath of consumers who may come into a given Buffalo Exchange. Moreover, each item is one-of-a-kind in the sense that the same exact item (by style, size, and color) has a low possibility of surfacing twice within the same location. Compounded by the ebbs and flows of the persons and goods that circulate in and out of the space, though still embedded within the larger fashion system of high street trends and neighborhood preferences, the Buffalo Exchange cannot place a premium on one aesthetic value over another. The same can be said for a close derivative, brand value. Therefore, many times, sellers who are accustomed to the Fashion Market are often shocked that their X or Y branded clothing is cast off by the discerning eyes of Buffalo Exchange buyers, as typified by this overheard conversation between a seller and the buyer:

> These are all AG! [Adriano Goldschmied]!
>
> —frazzled seller
>
> We don't buy based on labels, we buy based on style. We're looking for styles that aren't really well represented right now.
>
> —calm and collected Buffalo Exchange manager/buyer

In other words, reselling value at Buffalo Exchange is anything but the objec-tive 50% off MSRP warehouse price at which manufacturers would sell to Fashion distributors such as Saks Fifth Avenue or to its own employees. Reselling value is highly subjective to what the buyer deems as a garment's value. Rules governing the pricing are relative to that particular store's style needs, visible wear of the item,

---

[4]This is so ubiquitous that notwithstanding my being an avid patron of Buffalo Exchange, I was not even granted an interview when I applied for a side gig at the Allston location after college graduation.

temporal season, and weather. Brand value is a marginal consideration, unless the brand is a well-known runway favorite or a brand generally loved at the chain.[5] Even still, within the last couple of years, with my accumulated store credit from selling my unwanted commodities, I have been able to score a Richard Chai silk overcoat at about $65, an oversized Vivienne Tam silk and wool vest for $50, and a pair of Dolce & Gabbana wool slacks for $28 from the East Village and Chelsea locations. On several occasions during which I sold premium labeled denim, the store manager would nudge the buyer-in-training to up the resale value of my J Brand or Joe's jeans to the mid-thirties rather the high-teens to mid-twenties range of most of its denim. Sometimes, when the buyer realizes the quality of the material (e.g., silk rather than polyester), she will bump up the resale retail price a few dollars. Lastly, despite the fact that I have often sold items of clothing that still had its original tags attached to them, Buffalo Exchange buyers, echoing Kawanishi's curatorial statement, seemed largely immune to the additional value I hoped that the tags would imbue. The M-C-M' model need not apply.

Therefore, in the secondhand market of Buffalo Exchange, (un)correlative perspectives on value and "exchange value," viz. prices, complicate traditional Fashion System dichotomies of being materialistic. At Buffalo Exchange, individual sellers bring their preloved items that are no longer circulating within the Fashion Market to the secondhand markets. After each sale, the seller can choose to exchange her commodities for cold, hard cash (C-M) at 25% of the resale retail price or for store credit (C-M-C) at 50% of the resale retail price. Rather than following the M-C-M' circulation that is typical of the Fashion System, via its alternative value exchange system, the Buffalo Exchange encourages a C-M-C circulation in which the money circuit ends in a commodity that is to be purchased secondhand from within the Buffalo Exchange system. To that end, the Buffalo Exchange system encourages the closing of the loop in ways that the Fashion System does not. In the C-M-C system, the circuit begins and ends with the satisfaction of wants through consumption. With my store credit, I hardly ever use my own non-store credit money to buy from Buffalo Exchange. Thus, the loop is closed as the cycle begins and ends with commodities, not money. To wit, the value that is stabilized through immaterial desire and material consumption is use-value (Marx 2015 [1867]: 105), in this manner:

$$A(\text{no use - value}) = M(\text{exchange value/store credit}) = B(\text{use - value})$$

*where B's use-value equals or is less than the stored value in credit

In narrative form, I no longer have a use for commodity A and, I no longer value it. I exchange commodity A for value as store credit for use to consume commodity B, which materially holds a use-value that I immaterially desire.

---

[5]"Brands We Love": Abercrombie and Fitch, Universal Standard, Lane Bryant, Johnny Was, Kate Spade, Anthropologie, Bloomingdale's, Urban Outfitters, Ralph Lauren, Eileen Fisher, Bryn Walker, Marc Jacobs, The North Face, Barney's NY, Lululemon, Free People, Lucky Brand, Banana Republic, Topshop, Rag & Bone, ASOS Curve, Michael Kors, Nordstrom, Patagonia, Calvin Klein, ModCloth, Cole Haan, Madewell, Loft, Gap, Zara, Torrid, ASOS, AG, Vince, Nike, PAIGE, Adidas, Eloquii, J.Crew, Zara Man, Topman, Supreme, Coach, Vans, Boden, Stussy, Levi's.

4 Des-Vestire: Social Divestment Protocols and Valuing Socially Necessary Labor Time

In terms of financial markets and economics, divestment means the selling of financial assets, usually with the connotation of the sale being for some kind of ethical or political ends. Whereas in the financial landscape divestments tend to start phrases and sentences that end in "from fossil fuels," in a dress practices understanding, however, divestment is something that anyone who wears clothes performs on a regular basis. Divestment is necessary: We all let go of some clothes as our tastes change and as our bodies change. If there was a fashion equivalent of fossil fuels, it would be fast fashion chains such as Forever 21 and H&M. Despite the extensive scholarship, media attention, and moral outrage over the labor conditions at factories that produce fast fashion and the resulting environmental damage from mass production practices, Western modern liquid societies feed off the speed of these fashion flows. Engagement in the secondhand fashion market, such as Buffalo Exchange, is seen as a slow fashion tactic to combat dominant fast fashion practices, and there has been extensive literature in fashion studies that explore the consumption of secondhand fashion and often in "Third World" contexts (cf. Karen Tranberg Hansen on the *salaula* in Zambia 1995). However, the divestment procedures that exist in the (socially necessary) labor processes of producing these wares for the secondhand market has not been fully fleshed out.

Mobilizing my auto-ethnography, I perform wardrobe divestment rituals around every two months with each major turn in weather. Over the last three years, I have been gradually reducing the amount of color in my closet so that I can get to as close to a capsule wardrobe as I can. Each time I perform my divestment ritual, I am realizing this point by Marx: "A commodity, in its capacity of a use-value, satisfies a particular want, and is a particular element of material wealth. But the value of a commodity measures the degree of its attraction for all other elements of material wealth, and therefore measures the social wealth of its owner" (Marx 2015 [1867]: 86). As I think of the items that I will divest from my closet, I am thinking along similar terms as does an investment manager: Which stocks are not performing well? What are the trends for this particular industry? These similar forecasting questions are internally asked and answered as each item is scrutinized along corresponding criteria: Is this "me"? Do I see myself wearing this a few months from now? How much wear do I actually get out of this item?

In his critique on Marx by using the conceit of the thinker's coat, literary theorist Peter Stallybrass attributes another type of value, other than the (re)selling value, of an object. He claims that objects hold memories, which imbue upon the object sentimental value. Moreover, in order to put the object on the secondhand market, e.g., Buffalo Exchange, an object must be stripped of its sentimental value so that its exchange value can lay bare (Stallybrass 1998: 195). Memories, both in the metaphysical sense and in the tailoring profession's vernacular for "wrinkles," establishes the relationship between the body and object. Because metaphysical and literal memories devalue the exchange value of my commodities for the Buffalo Exchange buyer, wares to be exchanged in the secondhand market must be neutralized of literal wear and embedded memories (Stallybrass 1998: 196). This need to divest from your wares is also emphasized in promotional literature found in Buffalo Exchange stores and its corporate website.

As part of my own divestment rituals, I endeavor to dewrinkle my clothes as much as possible in both the literal and metaphysical senses. I wash my clothes to make them clean and smell nice. I take care to peel off the layers of red stickers on the clearance tags of the clothes that I had purchased on the adrenaline of the bargain hunt but have never actually worn. Sometimes, I purchase lightly worn to like-new items from thrift stores such as Goodwill for rock bottom prices and then resell them at the Buffalo Exchange for a net profit. With all of the clothes I take to the secondhand market, I snip off loose threads and button every button. Packing away the associated memories of wearing the particular garments, I delicately lay each folded garment onto neat stacks in my reusable Buffalo Exchange bags. Lastly, I sling the bags over my shoulder and make the twenty-minute trip to the East Village Buffalo Exchange site.

This begs another question: how do divestment rituals change the value of the commodity? Stallybrass answers it simply: the immaterial value of memory decreases the material value (resale retail price) that can be commanded for a given object. However, not so dissimilar from Stallybrass's analytical context of the pawnshop, the secondhand clothing market requires that wares produced for sale be as dewrinkled as possible. Yet, a balance of narrative fodder lubricates the social exchange mechanisms. In a sale in early spring 2017, I brought many of my brighter colored clothes that I now find to be completely "not me." In truth, many of those orange and pink garments had been purchased to appease my mother who longed for me to look more marriage material. A particular trapeze midi-dress from Anthropologie represents the antebellum Southern belle I could never really be. The last time I wore that dress was to the first orientation day of the Parsons M.A. Fashion Studies program, when I cared more to mind appearances that I looked like I cared (enough) about Fashion. Most of the other items in the orange and pink stack I have never worn. In casting off this self that was never "me," I joked with the Buffalo Exchange buyer, "Yeah, these are great pieces, but I'm just *too yellow* for them!" By caricaturizing my own Asian race and skin color, I provided narrative fodder that expedited the material/immaterial social exchange and secured the sale of those items. As Emily Spivack has shown in her earlier curatorial work, "Sentimental Value" (2007–2014), and as Ken Hillis (2006) has elaborated with his study on the language in eBay selling posts, the right combination of personal narrative and associative comedy ironically adds value to the sale of secondhand goods. Just as the selling of eBay items occurs on the open Web, the exchange of secondhand goods at Buffalo Exchange exchanges occur *en plein air* storefront counters (Fig. 2).

Another response to the above question is to re-conceptualize time in divestment as (socially necessary) labor time. At first blush, the labor time—viz. washing, snipping loose threads, taking clothes to the market for exchange, and creating relationships with buyers—in producing items for the secondhand market does not factor into the exchange value of the commodity. After all, in the exchange, Buffalo Exchange

Fig. 2 Anthropologie dress (MSRP $300+, bought at around $25, sold to Buffalo Exchange for $16.50) that was never really "me" and that I self-declare to be "too yellow" to wear

buyers are not paying for my time spent in making my wares suitable for the market. Yet, from my observations of the exchange at play between various buyers and sellers, the sellers who haphazardly shove their clothes into garbage bags and haul them by the trunkfuls to the market often fare less well in their sales. Now, with more than six years of experience in selling to Buffalo Exchange, I expedite the labor time the buyer spends in going through my wares. By already pre-curating clothes that I know are out of seasonal weather or are too worn and reconciling them to a donate-to-Goodwill pile and by performing my divestment protocols, I produce commodities for the market that can be quickly valued for resale. In a way, I increase my labor to reduce the labor of Buffalo Exchange buyers: They spend less time per item scrutinizing the details and less total time on my wares overall. Yet, at all stages of divestment—from relinquishing memories and former selves to creating narrative to oil the personal interactions between seller and buyer—the social aspect of production is valued. The cleaner, better materially kept my wares are, the more my wares are valued. The more I can socially lubricate the transaction between me and the seller, the more fruitful the sale will be.

5   Hoarding, Stored Value, and Closed Circuits of Exchange

Because the monetary value of the trade of my commodities for store credit is the more favorable deal at Buffalo Exchange, I hoard value on my trade card. At time of my writing this chapter, I had around $450 on my two trade cards. At my selling apogee, I had over $1000 in hoarded trade value. Marx's criticism of the hoarding of money is scathing: "This antagonism between the quantitative limits of money and its qualitative boundlessness, continually acts as a spur to the hoarder in his Sisyphus-like labor of accumulating (Marx 2015 [1867]: 86). In the analog of hoarding as a Sisyphean task, particularly in the context of Fashion, there is a direct link to Bauman's critique that Fashion consumption exists on a perpetuum mobile. Yet, what is out of relevance in Marx's critique is that liquid modern societies no longer materially store money as gold bars in bank vaults. In the way that we liquid modern industrialists use credit cards and spend money that we do not tangibly have, in the Buffalo Exchange system, the material embodiment of expired use-value turns into immaterial exchange value, as follows:

ITEM	MSRP ($)	BOUGHT ($)	SOLD ($)	CREDIT ($)	NET ($)
Diesel gray distressed skinny jeans	$198-$298	$4.49*	$32.00	$16.00	$11.51
J Brand blue knee patch skinny jeans	$188-$248	$12.00*	$36.00	$18.00	$6.00
Betsy Johnson gray sweatshirt	$80-$90	$12.00	$20.00	$10.00	($2.00)
Vera Bradley zip wallet/ID holder	$24.00	~$12.00	$16.00	$8.00	($4.00)
Club Monaco NWT printed blazer	$239.00	~$30.00	$36.00	$18.00	($12.00)
Club Monaco NWT cotton linen blazer	$249.00	~$30.00	$32.00	$16.00	($14.00)
Anthropologie trapeze printed dress	$300+	~$25.00	$16.50	$8.25	($16.75)
Banana Republic mini skirt	$80-90	~$7.00	$20.00	$10.00	$3.00
Carmar NWT grommet shorts	$228.00	$24.00	$28.00	$14.00	($10.00)
Dooney & Bourke small crossbody	$190-$210	~$30.00	$35.00	$17.50	($12.50)
Madewell split back Oxford t-shirt	$58.00	~$18.00	$17.50	$8.75	($9.25)
Seychelles black suede mules	$110.00	$10.00	$28.00	$14.00	$4.00

The preceding chart brings the exchange altogether. It shows the M′ of MSRPs of commodities for sale in the Fashion System and its value (clearance price) reduction in clearance sales or secondhand markets where I would pick up the item at a price befitting my ascribed use-value for it. The used commodities then become divested and sold in exchange for store credit to be used for future purchases of commodities that have use-value to me. Then, as highlighted in yellow, I show that the cycle can be activated again by re-introducing commodities purchased from secondhand markets into the Buffalo Exchange system. Within the sample set I provided, one item—the Diesel jeans—was purchased recently from an East Village charity thrift store. After my divestment protocols, the item was re-injected into the secondhand market, experiencing a change in value (resale retail price) that resulted in net profits for me. Similarly, the J Brand jeans were purchased in the last few months from the Buffalo Exchange Chelsea location at half off its initial resale retail price ($24) and then exchanged for store credit ($18) for it to be re-resold at $36 at the chain's East Village location.

Up until fall 2016, Buffalo Exchange always reduced the price of items that had stayed in its location for longer than a month to half off its initial resale retail price. Now, sale items are 25% off the initial resale retail price during the first two weeks of each month and then 50% off the initial resale retail price during the latter two weeks of the month. To that end, when prices at Buffalo Exchange gets marked at half off, then the C-M-C relationship for me the seller becomes effectively C-C in an equal exchange of store credit (50% of the selling item's resale retail price) to the new commodity's priced value (50% of the bought item's initial resale retail price). At this point, any systematic approach to determining M exchange value is overburdened with desynchrony at all stages of the exchange, resulting in what I am terming as M*.

By looking at the hoarding processes in my engagement with the Buffalo Exchange system, I have shown that hoarding occurs through my divestment rituals of collecting and preparing the commodities for injection into the secondhand market. At any given time, there may be multiple bags in my apartment containing items categorized for the resale market and for Goodwill donation. As Marx does consider the hoarding of commodities to be "sheer tomfoolery," the commodities do not remain in a hoard but become released in an extra-Fashion System secondhand market that (re)cycles the consumption and use-value life of the commodities (Marx 2015 [1867]: 415). Hoarding only happens within a temporal limit—that is, until the hoarded store credit is used to engage the seller as personal buyer in the secondhand market. In investment terms, the seller engages in new stock (clothing) picks to invest in herself.

6   In-Vestire: Taking Stock in Clothing

I paid $370 for a pair of Saint Laurent heels, but they were originally $960 [...] I NEVER — and I mean NEVER — pay full price for designer clothes! I always wait until it goes on sale or find a way to get it at a discount price. I am a serious thrifter, and I search out the

best of the best designer at Goodwill, Buffalo Exchange, Salvation Army, eBay, or I'll barter pieces with friends.
—Hannah, 31

I would spend $3,400 to $5,100 on the Chanel Boy bag. The Chanel Boy bag has been slowly replacing the Chanel 2.55 and seems to be sturdier. Its resale value is great right now (80% of original cost) so, the second the tides start turning against [the style], I'd sell it.
—Tina, 26

The preceding quotations, excerpted from a *Refinery 29* article on millennial women's biggest splurges, provide a snapshot of my contemporary peers' consumption attitudes toward big-ticketed items (Chou 2016). Often, luxury labels (e.g., Chanel, Burberry, Céline, and Chloé), "timeless," "classic" silhouettes (e.g., trenchcoats), and "wardrobe staples" (e.g., leather jackets) are invoked as investment pieces for these women. Calculations of cost per wear are made based on general factors of the longetivity of the style (aesthetic value), personal fit and feel-goodness (aesthetic value), durability (material value), and designer reputation (brand value).

With the sum of the money I "made" in one previous sale ($158.50), I reinvested the hoarded store credit into "investment" pieces as described by my millennial peers (see chart below, grand total of $155.50 with $3 to spare!). The chart is as follows:

Exchange (value) item	MSRP ($)	Bought ($)
Fendi black leather flats	$400+	$22.50*
Nicole Miller Atelier mixed media (leather and fabric) pants	$550+	$60.00
Vince black leather leggings	$995	$45.00
Etro polyester metallic jacket	$1100+	$28.00
		$155.50

Without running a regression, it is evident from the above sample set, the setting of resale retail prices is also clearly not dollar-by-dollar correlative to, or valued for, the original MSRP of any given item. As shown with the specific purchase of the Fendi flats, the C-C relationship exists for the items I sold, for which I then received store credit, and then used to purchase something that has greater use-value to me. From a net quantity in terms of items of clothing, I also have fewer pieces. Yet, the clothes I have received in exchange from the secondhand market are generally made from more durable materials (e.g., leather), feature more sophisticated construction, and promise greater quality so that I can get more cost per wear from them. Moreover, conscious of the future self that I am trying to fashion, I am actively investing in (in)vestments for a higher, better "self" constructed by wardrobe staples, not ephemera.

7   The Impactful (in)Vestments of a New Materialist

New materialism builds upon Marx's historical materialism in a way that can account for other values and powers of objects. More importantly, it unveils the social relationships, not just labor, that is hidden in objects. Well encapsulated by sociologists Nicholas Fox and Pam Alldred, "new materialist scholars assert that matter [should be] studied not in terms of what it is, but in terms of what it does: what associations it makes, what capacities it has to affect its relations or to be affected by them, what consequences derive from these interactions" (2017: 24). Moreover, new materialism offers ways to unravel how objects are indeed what Marx described as "social hieroglyphic[s]" (2015 [1867]: 49). New materialism can answer how objects can make people orient themselves differently and (dis)empower them, viz. thing-power. In short, new materialism uncovers the various concealed social relations between people and objects and between objects themselves.

To highlight the affective discordances in historical materialist critiques of Fashion and the Fashion System, cultural theorist Maurizia Boscaglia summarizes and contextualizes historical Marxist critique diffractively with new materialism:

> In a culture of abundance, such as consumer culture, the 'use' of clothes is always unnatural, for it is contra use value…The syntax of this unnaturalness is fixed in fashion. When we talk of clothes, and of the unnatural use of clothes, we immediately invoke an opposition between function and ornament, fact and fiction, reality and fantasy, materiality and abstraction—a split that in the west has been deeply gendered, so that the useless has been naturalized into the artificial, while the unproductive pleasure of a "decorative" female spectacle has been represented and perceived as illusion (2014: 93).

While Boscaglia helps to make a feminist claim to study clothing and, in her analysis, female characters' use of "useless" clothing to move plot points, I take her greater point to be that clothing when on the body can move us in ways that are overlooked as "useless" (cf. Ahmed 2010; Bennett 2004).

A beloved navy and pink floral Ann Taylor pantsuit has allowed me full social access to privileged places such as the midtown Manhattan law firm where I spend my time outside the academy. Yet, wearing this same suit makes me feel out-of-place in the more youthful and industrial spaces of my Brooklyn neighborhood. In spring 2017, at an international conference in France, I brought out that suit again so that I can appear "suitable" to give my presentation on the power dynamics embedded in female-to-male cross-dressing. In short, the social relationships inherent between myself and my suit as well as between myself and the people with whom I interacted was mediated through my suit. It was not mediated through the materiality or production processes of the suit, but through the encoded social ties that bounded my access, belonging, and displacement as a clothed body. Moreover, my arsenal of leather, tweed, and silk pieces purchased from secondhand markets are the (in)vestments I use to fashion my body and underwrite my own valuation of my professional self in the privileged academy.

8 Closing the Loop on the Wardrobe Gap: The Example of Brass

In an interview with *Digiday*, Brass Co-Founder Katie Demo summarizes this sartorial endemic facing women: "I really think it's a deep societally ingrained thing. There's this sense that, with men, it doesn't matter what they're wearing. Their ideas stand for themselves… For women, it's your appearance plus your ideas. For men, those things are separate" (Biron 2017). Founded in 2014, Brass's founders Katie Demo and Jay Adams are seeking to close what they have termed the wardrobe gap for women. Or, within the understanding of this analysis, they are addressing the gap in (in)vestment opportunities for women who seek to raise their own valuation in the gendered workplace.

Since their founding of Brass, Demo and Adams have surveyed and spoken to over a thousand members of their customer base to elucidate the following stark reality of the wardrobe gap for women: "77% of our customer base believes the way they look directly influences their success at work… 84% of our customer base have negative to neutral feeling about their wardrobes… [and of] the women we surveyed said they feel good about their appearance only 50% of the time" (Brass 2017). Adams sheds further light on their survey results:

> 82% of women went through a significant change in the last 2 years (new job, moved to a new city, got married, had a baby). 77% of these women were in the mind set to pare down their wardrobe and/or start building a foundational wardrobe and change their shopping habits. Our take-away from this was that women closely link their personal appearance with external situations. It could be argued that when women receive external validation of their 'worthiness' they make investments in their personal appearances. But this external validation must come first thereby providing women with the feeling that they 'deserve' something (2017).

As entrepreneurs, Demo and Adams funded this project in a way that is as unorthodox as their concept. Rather than choosing to appeal to generally male venture capitalists or to friends and family for a loan, Demo and Adams choose to crowdsource on Kickstarter, ultimately raising over $27,000 to launch their first spread of five effortless dresses in early 2015 (Brass 2015). Here, this analysis concedes a rhetorical concept: that for entrepreneurial efforts to be mainstream sustainable often functions a luxury that most female founders do not have. As Adams illuminates:

> There are definitely fewer funding options available to women and to consumer product/retail. Actually, I should rephrase this. It is not that there are fewer options, it just happens that women in general receive a very small portion of VC money (something around 4%, I believe). There are also few examples of women-founded and led fashion brands that have grown to be $100+million in annual revenue. The fewer examples of success stories, the harder it is to paint the picture of success to an investor. We are starting these conversations now and are taking a more "masculine" approach. Men have an incredible way of feeling worthy of investment right out of the gate. Women, on the other hand, tend to feel that they should "prove" themselves first and then they will be worthy. This attitude does not serve you when asking an investor for $1 million to fund the potential of an idea (2017).

Speaking at the 2017 Hacking Arts Conference at the Massachusetts Institute of Technology, Adams also made explicit the importance of truly closing the loop in a circular economy for fashion items. Brass operates on the Brass Rule: "Treat your customer how you want to be treated. And never lie about price." In fact, Brass is so transparent that it offers a "Behind the Seams" blog series that documents the design process and consideration of every major item. Contra ethical fashion brands such as Reformation and Everlane, Brass does not position itself in the market as an ethical fashion brand, per se, despite their use of quality materials that have lower environmental impact as well as their manufacturing in factories with equitable labor practices. And unlike MM. LaFleur, Brass does not price its items at M' values that create financial barriers to access for many younger professional working women. Moreover, Brass further reduces its carbon footprint and costs by keeping its experience mostly online with the exception of one Style Studio, near their Boston headquarters, where women can bring their own clothing to get styling tips.

It is clear that Brass is passionate about closing the loop. Their signature Closet Kit of pick-your-own three wardrobe staples comes with a Clean Out Bag for customers to glean the clothes that they do not need and send it back to the company so that it can send it to a clothing recycling facility. In addition, according to Adams, Brass is working toward a business model of buying back used Brass items from customers to resell at a lower price point. This model further encourages the closing of the loop by providing divestment tools and avenues for consumers to responsibly release and then inject fashion items back into the fashion cycle.

9 Conclusion

As with the rewards of investing in real stock, the real (in)vestment of clothing through purchase and embodiment on bodies provides insights that material clothing can provide its human wearers certain affordances (Kaptelinin n.d.). As evidenced in the responses of my peer millennial women and of Brass clients in thinking about investing in themselves, they seek to take stock in the items that will generate returns for them in the long run. They invest or hope to invest in items that will give them the additional confidence boost in spaces in which they feel alienated or that will allow them to become a better, future self. By analyzing the ideas of investment and divestment diffractively through its etymological root of *vestire* (to clothe) and auto-ethnography of my secondhand clothing practices, I hope to "disrupt linear and fixed causalities, and to work toward 'more promising interference pattern'" in attitudes about fashion outside of the Fashion System (van der Tuin 2011: 26). I show that in secondhand fashion markets, social labor production of products is out in the open, literally at the storefront counters of the Buffalo Exchange, and the cycle of values is not linearly correspondent to liquid Western capitalistic market exchanges. Furthermore, by providing the example of Brass, I assert that such diffractive opportunities are also possible in current extra-industry endeavors so that the luxury of presenting a professional self is no longer out of reach.

After all, my suit moves me. It materializes a better version of me. Not only do we millennial women materially invest in our clothing and our image, often using calculations of alternative and immaterial values hidden and readily dismissed by those who think fashion is "useless," but we also (in)vest our bodies in our clothing. Ergo, impact (in)vestments. When I put on a suit, my body makes a contract with my suit. The squared shoulders of the blazer pull my physical shoulders back erect and straighten my spine: I stand a little taller. Yet, my arms fall naturally akimbo. The single button front of the blazer rests at my center core, reminding me to create a direct line from my lifted chin to my core. The pants' waist sits right above my hips, moving them to widen to form a solid base. The hem of the pants taper to hug my ankles and beg to be met with an elegant heel. I oblige with a coordinating pair of navy leather peep-toe pumps. The suit orients me to stand in a Wonder Woman pose and walk with a sense of purpose. And most poignantly for me—a queer, immigrant, woman of color—in my suit, I value myself as someone worthy to be in the privileged spaces and hallowed grounds of the academy. If I can look the part, then maybe I can claim a seat at the table. My suit *suits* me.

My clothes move me. They move me to social divestment protocols that allow me to materially and immaterially shed former selves or selves that never were really "me." This pile of stuff—silks, jacquard wools, suedes, cashmeres, and cottons—orients me to take stock in material clothing, invest in myself, and raise my valuation. Thus, this analysis makes the seemingly contradictory conclusion: Value itself is complicated and nebulous yet the immaterial valuation of ourselves is the total project of consuming materially. By explicating (in)vestments, the impact of this seemingly trite phrase gains material evidence: "If I look good, then I feel good." And by showing the interferences of alternative conceptions of value in historical materialism when applied to secondhand clothing markets, I ultimately articulate a reclamation for new materialism and fashion studies. That is, fashion, with a lower case "f."

References

Adams, J. (2017) Email correspondence with KY Deng. Retrieved from November 27, 2017.

Ahmed, S. (2010). Orientations matter. In D. Coole & S. Frost (Eds.), *New materialisms: Ontology, agency, and politics* (pp. 234–258). Durham: Duke University Press.

Bauman, Z. (2006) Liquid fear. Polity, Cambridge.

Bauman, Z. (2010). Perpetuum mobile. *Critical Studies in Fashion and Beauty, 1*(1), 55–63.

Bennett, J. (2004). The force of things: steps toward an ecology of matter. *Political Theory, 32*(3), 347–372.

Biron B (2017) 'The wardrobe gap': clothing brand Brass is out to change officewear for women. Digiday. https://digiday.com/marketing/wardrobe-gap-clothing-brand-brass-change-officewear-women/. Retrieved from November 13, 2017.

Brass (2015) 3 reasons we chose Kickstarter over venture capital. Medium. https://medium.com/@BrassClo/3-reasons-we-chose-kickstarter-over-venture-capital-ec1d899980b7. Retrieved from November 13, 2017.

Brass (2017) The wardrobe gap. Medium. https://medium.com/the-workroom-by-brass/the-wardrobe-gap-d50dc1e7a837. Retrieved from November 13, 2017.

Boscagli, M. (2014). *Stuff theory: everyday objects, radical materialism*. New York: Bloomsbury.

Chou J (2016) I spent $2,580 on a purse: 30 women on their biggest fashion buy. Refinery 29. http://www.refinery29.com/investment-clothing-splurge-buys. Retrieved from March 12, 2017.

Clark H, Laamanen I (2017) Fashion after fashion. Museum of Arts and Design. New York, New York, April 27–August 6, 2017.

Fox, N. J., & Alldred, P. (2017). *Sociology and the new materialism: theory, research, action*. London and Thousand Oaks: Sage.

Hansen KT (1995) The world of Salaula: Clothing and value contests in Zambia. Institute for Advanced Social Research, University of the Witwatersrand, Johannesburg.

Hillis, K. (2006). Auctioning the authentic: eBay, narrative effect, and the superfluidity of memory. In K. Hillis, M. Petit, & N. S. Epley (Eds.), *Everyday eBay: culture, collecting, and desire* (pp. 167–184). New York: Routledge.

Kaptelinin V (n.d.) Affordances. In: M. Soegaard & R.F. Dam (Eds.), *The encyclopedia of human-computer interaction*, 2nd edn. Interaction Design Foundation, Aarhus.

Manhattan East Village. Buffalo Exchange. https://www.buffaloexchange.com/locations/new-york-city/manhattan-east-village/. Retrieved from March 17, 2017.

Marx, K. (2015). *[1867]) Capital* (Vol. 1). Moscow: Progress Publishers.

Spivack E (2014) Sentimental value. http://www.sentimental-value.com/. Retrieved from May 8, 2017

Stallybrass, P. (1998). Marx's coat. In P. Spyer (Ed.), *Border fetishisms: Material objects in unstable spaces* (pp. 183–207). New York: Routledge.

Van der Tuin, I. (2011). A different starting point, a different metaphysics: Reading Bergson and Barad diffractively. *Hypatia, 26*(1), 22–42.

Responsible Luxury Development: A Study on Luxury Companies' CSR, Circular Economy, and Entrepreneurship

Carmela Donato, Cesare Amatulli and Matteo De Angelis

Abstract In this chapter, we discuss how luxury brands can build their success on corporate social responsibility (CSR), leveraging specifically on the paradigm of circular economy. The idea advanced in the chapter is that luxury and sustainability are not conflicting concepts, as many believe, but they are positively correlated, inasmuch as the quintessential characteristics of luxury goods make them potentially more sustainable than mass-market goods. Through the discussion of four case studies of luxury brands operating in the sectors of fashion (Brunello Cucinelli, Gucci and Stella McCartney brands) and food (Godiva), we point out that the reuse of tangible resources, such as money generated by companies' activities and raw material, can be a very solid basis for building market success as well as to broaden the positive contribution luxury brands can make to the environment, the employees, the local community of producers, and, as a consequence, to the society at large. A common feature of all the cases discussed is represented by the key role played by the entrepreneur (often the founder of the company) in fostering the balance between brand prestige and sustainability.

Keywords Luxury · CSR · Sustainability · Circular economy
Entrepreneurship · Environment · Employees · Society

C. Donato (✉) · M. De Angelis
Department of Business and Management, LUISS University, Viale Romania 32,
Rome 00197, Italy
e-mail: donatoc@luiss.it

M. De Angelis
e-mail: mdeangelis@luiss.it

C. Amatulli
Ionian Department in "Legal and Economic Systems of Mediterranean: Society, Environment,
Culture", University of Bari, Via Duomo 259, Taranto 74123, Italy
e-mail: cesare.amatulli@uniba.it

© Springer Nature Singapore Pte Ltd. 2019
M. A. Gardetti and S. S. Muthu (eds.), *Sustainable Luxury*,
Environmental Footprints and Eco-design of Products and Processes,
https://doi.org/10.1007/978-981-13-0623-5_2

1 Communicating Sustainability and CSR

Sustainability is certainly one of the key determinants of future economic and social development. Both scientific research (e.g., Nidumolu et al. 2009) and business reports (e.g., Deloitte Development 2011) agree that sustainability has become a crucial driver of innovation in several business sectors; as such, it serves as a critical success factor for firms. In light of this evidence, companies' competitiveness will be increasingly based on their ability to orient (in some cases) or re-orient (in other cases) their overall vision, mission, and strategy—and by extension, their product, branding, and communication activities—to sustainability.

In this chapter, we focus on sustainability-oriented communication strategies and tactics that might aid companies in terms of improving consumers' attitudes and increasing their purchasing intentions. More specifically, we analyze sustainability-oriented marketing and communication initiatives in one sector that, as we will explain in greater detail below, is often considered to be quite antithetic to sustainability—luxury. There is great value in understanding the way(s) in which sustainability initiatives should be designed and communicated to garner consumers' favor. However, scientific research has devoted scant attention to the factors that drive the effectiveness of communication strategies aimed at promoting sustainable products (e.g., advertisement messages delivered across different media outlets or messages conveyed by salespersons).

According to The Brundtland Commission,[1] *sustainable development* is "the ability to … [meet] the [needs] of the present without compromising the ability of future generations to meet their own needs". This definition is clear, but also general enough to encompass different companies operating in different business sectors. Importantly, it also covers different aspects of sustainability, referring to not only the preservation of the environment, but also the well-being of communities, their stakeholders, and the society at large. This accords with recent academic research, which suggests that companies are paying greater attention to sustainability issues involving the environment, their employees, customers, and society at large (e.g., Gershoff and Frels 2015; Luo and Du 2015). A similar trend of rising interest can be seen among consumers and the general public, facilitated by their exposure to sustainability issues on the vast array of digital and social media platforms available today. For instance, consumers have become quite sensitive to how—and to what extent—companies dedicate their commercial profits to tangibly improving the well-being of relevant stakeholders, communities, and the surrounding environment.

Aware of the growing interest in sustainability development, companies are seeking to not only design sustainability actions, but also craft communication strategies and messages that can publicly reflect their ability to benefit the environment, their employees and other relevant stakeholders. Such sustainability-oriented initiatives typically fall under what is widely known as companies' Corporate Social Respon-

---

[1] In 1983, Brundtland was invited by then-United Nations Secretary-General Javier Pérez de Cuéllar to establish and chair the World Commission on Environment and Development (WCED), widely referred to as the Brundtland Commission.

sibility (hereafter CSR). CSR can be defined as a company's continuous effort and commitment toward behaving ethically and contributing to the improvement of the general quality of life (Maignan and Ferrell 2001). Amidst the mounting pressure of mass-media coverage, consumer advocacy groups, and rising numbers of Web sites dedicated to company ethics (e.g., www.companyethics.com), companies perceive the urgency to engage in—and importantly, effectively communicate—CSR initiatives. Consistently, scholarly research highlights the strategic relevance of CSR for most companies, due to the high public exposure of their behavior (Bielak et al. 2007). Several studies show that information on companies' CSR actions affects consumers' attitude toward firms (e.g., Brown and Dacin 1997), their purchase behaviors (e.g., Mohr and Webb 2005) and the extent to which they identify with the company (e.g., Bhattacharya and Sen 2004). Companies, therefore, have tried to centralize sustainability-related activities by creating CSR departments. These departments are generally tasked with managing and communicating a company's commitment to minimizing the harmful effects of its business while maximizing its long-term, positive effects on society at large (Mohr et al. 2001). Since sustainability issues cross all areas of the business—from manufacturing to retailing, from transportation to human resource management—companies typically create CSR departments as a way to organize activities that are multi- and cross-department, while still allowing a firm to transcend mere legal compliance by generating positive social value (Alvarado-Herrera et al. 2015).

While CSR departments are becoming a common practice in several businesses, they are neither universal nor absolutely. Companies can still reach sustainability goals through other organizational approaches. According to Langenwater (2009), it is possible to distinguish three dimensions that reflect the essence of sustainable development: the *social* dimension, which assumes that sustainable development contributes to the general improvement of life quality; the *environmental* dimension, which assumes that sustainable development involves reducing natural resource usage while producing eco-friendly goods, and the *economic* dimension, which assumes that sustainable development involves the regulation of resources usage across generations. Considering that CSR is usually aimed at integrating social and environmental aspects into corporate activities (Baumgartner 2014), it is possible to claim that CSR and sustainability are deeply related concepts, most of the time used interchangeably, however whereas for business sustainability reflects their concerns for the long-term impact of their activities on future generations and the environment, CSR can be interpreted as a means by which companies formalize their commitment to sustainable development (D'Anolfo et al. 2017).

2 Circular Economy and CSR

One recent concept that has garnered substantial attention from business practice, and relates closely to sustainability, is circular economy (hereafter, CE). CE may be defined as a "regenerative system" in which resources and waste are minimized by slowing, closing, and narrowing material and energy loops (Geissdoerfer et al. 2017). In other words, a circular economy restores material flows through closed-loop processes, reusing valuable resources and creating less waste. In contrast to the so-called linear economy, CE encourages the co-existence of business development and environmental protection. Because of its circular (closed) flow of materials and reuse of energies and raw materials (see, for instance, Yuan et al. 2006), CE has become a central objective for institutions—and in particular, for the European Community through the European Circular Economy package (European Commission 2015). A clear and interesting example of a company that practically, and maybe unconsciously, follows the CE approach is the Italian startup Quagga, which has introduced a new type of sustainable winter jacket made of a new fiber produced from plastic bottles. By reusing other materials from a different industrial sector, Quagga's products reduce plastic waste in the environment. Moreover, when their lifecycle ends, those jackets are recovered and reused again for the production of new ones. In short, this new and small fashion company is able to build its branding strategy on sustainable values—and more specifically on the CE paradigm (Geissdoerfer et al. 2017)—without sacrificing aesthetics and design.

Naturally, many companies could be characterized in terms of their commitment to sustainability, CSR and CE. In this chapter, however, we focus on how such trends are shaping one particular industry: luxury. Indeed, the very idea of "sustainable luxury" represents an interesting knot to unravel (Kapferer 2010). It seems almost oxymoronic: luxury, with its echoes of hedonism, prestige, and excess, looks incompatible next to sustainability and its associated notions of sobriety, ethics, and moderation (see Davies et al. 2012). The prospect that luxury and sustainability might be synergistic is thus quite new to both scholarly research and managerial practice. As Amatulli et al. (2017a) highlighted in a recent book, such an idea is grounded on the evidence that luxury could be considered as *inherently* sustainable. Indeed, the core characteristics of luxury goods—their durability, craftsmanship and limited production—speak to the idea that luxury products and brands can have a positive impact on company employees, the environment, host territories, and the society at large. In order to legitimize this argument, luxury companies need to fully understand how to design marketing and communication initiatives that underscore the value of their sustainable actions.

Before that, though, it is necessary to highlight the need for a strong entrepreneurial spirit in a CE business model. This is especially true for luxury brands, given the continuous interplay between the managerial and the artistic "souls" that characterizes the luxury world. We understand entrepreneurship as the capacity to identify, evaluate, and exploit opportunities to create future goods and services that make a profit and are concerned with the discovery and exploitation of profitable opportunities (Shane and Venkataraman 2000). Luxury companies, even more than others, depend on entrepreneurship: They are typically founded by a designer or an artisan who develops the brand through a combination of style, beauty, production, and commercial skills. Case in point: The Giorgio Armani company, one of the most important ones in the luxury fashion sector, was founded in 1975 by the designer Giorgio Armani. A resident of Milan, Armani had the opportunity to develop a global brand through not only his exceptional and disruptive creativity, but also his entrepreneurial capability. Today, the Giorgio Armani S.p.A. is making important steps toward sustainable development thanks to its founder's personal initiatives and commitment to sustainability; indeed, it was Giorgio Armani himself who announced in 2016 that he would stop using fur for all Armani products.

Through a qualitative analysis of four luxury companies, the present chapter illustrates how, in the context of luxury brands, CSR may be linked to CE and entrepreneurship. In particular, each of the four case studies, we discuss the deals with initiatives belonging to one of the four dimensions of CSR (i.e., economic, legal, ethical, and philanthropic; see Carroll 1979, 1991). Through our analysis, we hope to stimulate luxury practitioners to think deeply about how to design and communicate initiatives that leverage the sustainability potential that is often intrinsic to luxury brands. Luxury companies may find this knowledge useful for developing a more structured CSR strategy that combines the CE approach with their managers/founders' transformational leadership behavior.

3   Luxury Brands and CSR Initiatives

Although many luxury brand managers have long understood that sustainability should be central to their businesses, they have largely neglected sustainability objectives. From a business perspective, of course, they did not need to be sustainable to do well on the market (e.g., Achabou and Dekhili 2013). In the wake of social media and increasing consumer awareness, however, many luxury companies have developed actions aimed at embracing sustainability and showcasing their commitment to environmental and/or social issues (see, for instance, Amatulli et al. 2017b). To mention some examples, Tiffany started certifying its diamonds as "conflict free", Chanel incorporated "earthy materials" in its 2016 collection, and Bulgari recently funded the restoration of Rome's Spanish Steps.

These examples suggest that luxury brands are interested in not only promoting their prestige and status, but also exhibiting altruistic and moral values through various sustainable practices. This in turn reflects a deep change in the typical luxury buyer, for whom egoistic motivations (e.g., communicating status to others) have become cushioned by altruistic motivations (e.g., preserving environmental resources). Driven by an increasing interest in the provenance and environmental impact of their luxury shopping (Lochard and Murat 2011), these new luxury customers enforce higher and clearer expectations that imbue luxury goods with a higher aspirational value. In this way, customers can not only project a positive image in social contexts and/or fulfill their personal tastes, but they can also feel like they contribute positively to society. As a consequence, luxury brand development in the current market demands a discussion about sustainability.

Table 1 Carroll's multidimensional CSR model

CSR dimensions	Responsibilities	Main potential goals
Philanthropic	Behaving as good corporate citizens	Contributing to the community Improving quality of life
Ethical	Being ethical	Doing what is fair and moral Avoiding harm
Legal	Following the law	Providing goods and services that meet legal requirements
Economic	Making acceptable profits	Creating the basis for all the other responsibilities

Source Carroll (1991)

However, seizing the opportunities of sustainability begins with having innovative people, with a deep sensitivity toward environmental issues, who are motivated to disrupt the prevailing economic paradigm of linear economy—namely, the take—make–use–dispose process. In most cases of luxury companies, such people are the entrepreneurs/founders. Indeed, there are several cases of global luxury brands where the entrepreneurs are developing CE models through specific CSR activities in order to promote their brands as sustainable. However, many of these attempts lack a systematic approach by which promoted CE activities can be classified into well-defined CSR strategies. To address this gap, our study aims to illuminate the role of CE in luxury by proposing a series of real-world cases of luxury entrepreneurs developing different types of CSR initiatives. In particular, our chapter is rooted in the well-established idea that CSR is a multidimensional construct (D'Aprile and Mannarini 2012). We apply Carroll's (1979, 1991) widely used multidimensional model of CSR, which describes four main dimensions—economic, legal, ethical, and philanthropic—that correspond to company's four types of responsibilities. Carroll portrayed such dimensions as a pyramid, with the economic dimension, whereby companies should "make an acceptable profit" (Carroll 1991, p. 141), at the bottom and the philanthropic dimension, whereby companies should behave as good corporate citizens, at the top of the pyramid. The pyramidal model is summarized and described in Table 1.

The economic responsibility is driven by the profit motive, the primary incentive for entrepreneurship. Indeed, business organizations have historically been designed to provide goods and services to societal members in exchange for an acceptable profit. By implication, all other business responsibilities are dependent upon the firm's economic responsibility: The company should strive to maximize earning per share, to be as profitable as possible, in order to maintain a strong competitive position and a high level of operating efficiency. With no such conditions, it would be hard for companies to stay in business long enough to establish a market position where they can positively contribute to society's sustainability.

The legal responsibilities encompass a company's need to follow legal norms. Firms are expected to pursue their economic missions within the legal frameworks established by federal, state, and local governments. Indeed, this combination of legal and economic responsibilities underlies the free enterprise system that codifies the basic notions of fair operations. To this end, companies should become law-abiding corporate citizens, provide goods and services that meet legal requirements, and perform consistently within regulatory boundaries.

The next layer of the pyramid involves companies' ethical responsibility, which captures the standard norms or expectations for what is fair and moral in the minds of consumers, employees, shareholders, and the community. Differently from the more tangible legal responsibilities, ethical responsibilities refer to the principles of moral philosophy, including concepts such as justice, rights, and utilitarianism. This implies that companies should perform in line with societal expectations and ethical norms, should recognize and respect new ethical norms adopted by society and, more importantly, should recognize that ethical behaviors go beyond mere compliance with laws and regulations.

Finally, at the top of the pyramid lies philanthropic responsibility. Ideally, companies that participate in this dimension want to be good corporate citizens that actively engage in actions or programs aimed at promoting human welfare and goodwill. Philanthropic and ethical responsibilities are distinct in that the former are not expected in a moral sense: Even though communities may desire that firms contribute their money and time to humanitarian purposes, they do not regard the firms as unethical if they do not provide that kind of contribution. This implies that companies should participate in voluntary and charitable activities within their local communities, provide assistance to private and public educational institutions, and assist those projects that enhance a community's "quality of life".

In sum, Carroll's CSR framework (1979, 1991) expects companies to produce and sell products at a profit, while controlling production costs or improving working conditions (i.e., *economic* CSR dimension); comply with the requirements imposed by the legal system at play (i.e., *legal* CSR dimension); endorse principles of fairness and justice in their activities (i.e., *ethical* CSR dimension), and engage in voluntary actions that qualify them as "good corporate citizens" (i.e., *philanthropic* CSR dimension). On the basis of this multidimensional framework, and following Pino et al.'s (2016) recently disentangling of each dimension's effect, we argue the following: CSR dimensions can be classified by their visibility to consumers. Whereas initiatives in the legal and philanthropic domains are easily visible to and noticeable by consumers, those in the economic and ethical dimensions are not (at least not immediately). Building on such this distinction, we identify two categories of CSR dimensions: the "internal" dimensions category (which encompasses the economic and ethical dimensions) and the "external" dimensions category (which encompasses the legal and philanthropic dimensions).

4 Sustainable Luxury Development: The Role of CSR Initiatives, Circular Economy, and Entrepreneurship

The aforementioned dichotomy may be a useful "tool" for classifying and formalizing initiatives that underlie the CE approach, especially for luxury companies. In other words, we propose that CE activities can be classified in terms of the CSR category to which they belong. Moreover, we propose that this external versus internal dichotomy may be used to investigate and formalize entrepreneurial activities in luxury. Therefore, luxury companies that are beginning their journey toward sustainability could develop their CSR strategy by pairing the CE paradigm with their entrepreneurial capabilities. By understanding how to develop and communicate the different CSR dimensions, luxury marketing managers may be able to enhance their brands' sustainability while maintaining their entrepreneurial vision and profitability.

Amidst this advice, we recognize that luxury companies want to become less wasteful without sacrificing their products' exclusivity. In this sense, they need to develop a conceptual framework that not only links CSR with CE and entrepreneurship, but also tailors CSR communication to the right segment of consumers. Past research emphasizes that on luxury consumption generally follows two different, and sometimes opposing, approaches. On one hand, consumers may see luxury goods as status symbols and mainly buy them to showcase their social status; on the other hand, consumers may buy luxury goods as a way to satisfy their individual taste and style. The former approach, centered on social visibility, is understood as "external" luxury consumption; the latter approach, centered on personal style, is understood as "internal" consumption (Amatulli and Guido 2012; Amatulli et al. 2015; Han et al. 2010; Nueno and Quelch 1998). We contend that this consumption dichotomy may have relevant implications for a CSR framework. Indeed, customers who mainly see luxury as a means to position themselves within society and show off their success to others could be particularly attracted to CSR initiatives that are highly visible and easily recognizable by others (i.e., widely identifiable from outside the company). Oppositely, customers who see luxury as a means to owning products that align with their inner values, individual style, high-quality standards, and intimate emotional needs could be less interested in the visibility of CSR initiatives. In fact, that may have more appreciation for those elements/initiatives that are more "quiet" than "loud" (e.g., Amatulli et al. 2017b; Han et al. 2010).

Therefore, luxury companies should consider developing external (i.e., the legal and philanthropic ones) or internal (i.e., the ethical and economic dimensions) CSR dimensions based on whether their customer base exhibits an externalized or internalized approach to luxury. If the firm's customers focus on the products' high price and conveyed prestige, then external CSR dimensions should constitute the focus; if customers prefer the brand because of its quality, stylistic elements, and values, then management should emphasize the internal dimensions of CSR.

Given the above considerations, this chapter explores the opportunity to connect CSR initiative with CE and entrepreneurship in luxury. More specifically, our explorative and qualitative study sheds light on how luxury brands' initiatives, whether

Table 2 Luxury consumption dichotomy and CSR development

The luxury consumption dichotomy	Main consumption drivers	Main luxury brand/product elements	CSR dimension to be emphasized by the luxury company	Case studies	Examples of Entrepreneurship and CE initiatives
Internalized luxury	Individual style	Stylistic identity	Economic	Brunello Cucinelli	Redistribution of a part of the profits to employees
	Inner values	High quality	Ethical	Gucci	Introduction of recyclable packaging
Externalized luxury	Social status	Prestige	Legal	Stella McCartney	Use of the regenerated cashmere yarn Re.Verso™
	Visibility	High price	Philanthropic	Godiva	Limited-edition's revenues used to help children in cocoa-sourcing regions

Source Elaborated by the authors

related to CE or entrepreneurship, may be classified according to the internal or external CSR dimensions.

To this end, we conducted a qualitative case study analysis on CE initiatives undertaken by entrepreneurs of luxury brands. In particular, we present four cases of luxury brands characterized by a strong entrepreneurial spirit; in each case, we describe the CE initiatives in terms of the four abovementioned CSR dimensions. Specifically, the first two cases—Brunello Cucinelli and Gucci—portray CE activities that belong to the internal CSR dimensions (i.e., economic and ethical dimensions), while the other two cases—Stella McCartney and Godiva—portray CE activities that belong to the external CSR dimensions (i.e., legal and philanthropic dimensions) (see Table 2 for a summary). We gathered most of the information for these cases from the Web sites of the analyzed brands.

5   Entrepreneurship and CE Initiatives Based on Internal (Economic and Ethical) CSR Dimensions

5.1   The Case of Brunello Cucinelli S.p.A

Brunello Cucinelli S.p.A., founded in 1978 as a small Italian cashmere producer, is now one of the best known brands in the luxury and casual-chic fashion sectors, with boutiques located in large cities worldwide. The company's success is owed to its founder, Mr. Brunello Cucinelli, who amazed the market with the idea of coloring cashmere.

Since he was a young boy witnessing his father's work afflictions, Cucinelli has been dreaming of a type of work that respects man's moral and economic dignity. This aim is crucial to understand Cucinelli's personality and business success. Cucinelli claims to see business as not only the mere production of wealth, but as a way of turning respect for human dignity into a core pillar of capitalism.

In 1985, Cucinelli purchased a fourteenth-century castle in Perugia (Italy) that became his corporate headquarters. In 2000, with the objective of meeting the market's growing demand with adequate production facilities, he decided to acquire and refurbish an existing plant at the foot of the Solomeo hill. He transformed this location into an ideal venue for culture and art by building a Forum of the Art, including a library, a gymnasium, a theater, and an amphitheater—and in this way, contributing to the welfare of the local community. In 2012, he listed the company in the Milan Stock Exchange, with the aim of not only improving its financial results, but also spreading his ideals of "Humanistic Capitalism" to the broader world.

In 2015, Cucinelli unveiled the "Project for Beauty" initiative, which entailed creating three huge parks in the valley of Solomeo Hill—a reflection of Cucinelli's desire to preserve green land. Over the years, Mr. Cucinelli has obtained a staggering number of national and international awards for his "Neohumanistic Capitalism"; however, the recognitions that best represent his human accomplishments are: his honorary degree in Philosophy and Ethics of Human Relations, awarded by the University of Perugia; the Global Economy Prize, received from the eminent Kiel Institute for the world economy with the noble mention that he "personifies perfectly the figure of the Honorable Merchant"; and the Sustainability Award, recently given by the Green Carpet Fashion Awards in Milan.

Due to these efforts, the company is a living example of CE; it "reuses" many of the economic resources it generates. In fact, in 2012, after Brunello Cucinelli S.p.A. was successfully listed on the Italian Stock Exchange, Cucinelli distributed a big part of the year's net profit (5 million of Euro) to its 783 employees, each of whom received a Christmas bonus of 6,385 Euro. Cucinelli intended this bonus as a proper recognition of his employees' good work in helping the company's profits grow by 25%. This was quite an exceptional action, considering that, at the time, Italy was struggling to emerge from a deep crisis. In this context, Cucinelli was an encouraging example of excellent results arising from a sustainable entrepreneurial philosophy. This sustainable action constitutes a CE activity promoted by an entrepreneur, in that it favored the circular flow of resources inside the company, but it can also be classified as an economic CSR activity, because the company's profits were the reused resource.

5.2 The Case of Gucci

Gucci is an Italian luxury brand of fashion and leather goods, part of the Gucci Group. It was founded by Guggio Gucci in Florence in 1921 after he was inspired by the extravagant and elegant baggage he saw while working in a hotel in London. Initially, Gucci sold luggage imported from Germany and offered repair services on the side. Facing a shortage of imported leather, Gucci innovated by using new materials, such as canvas, as well as producing small leather goods, including wallets and belts. However, it was after World War II that Gucci started to become an internationally known luxury name, once such luminaries as Grace Kelly, Elizabeth Taylor, and Queen Elizabeth became Gucci patrons. By 1953, Gucci had opened his first American shop in New York.

After Guccio Gucci's death, the company passed on the reins to his son Aldo, who successfully turned the Gucci name into a global fashion brand, launching ready-to-wear lines for both men and women by 1974. Unfortunately, internal disputes between Aldo and his brother, Rodolfo Gucci, started to threaten the company's winning formula, until a dramatic crisis in 1993 led to the company being acquired by Investcorp. In July 1995, Investcorp led the company to be listed on both the New York Stock Exchange and the Euronext Amsterdam Stock Exchange (Moffett and Ramaswamy 2003).

During last decades, Gucci initiated several partnerships aiming to sustain charitable projects. One of the most famous projects is partnership with UNICEF, for which Gucci stores worldwide donate a percentage of the sales of special collections made specifically for UNICEF to support education, healthcare, protection, and clean water programs for children and orphans affected by HIV in sub-Saharan Africa. Between 2005 and 2010, Gucci donated over $7 million to UNICEF. In February 2013, the company launched the "Chime for Change" campaign, aiming at promoting and supporting the global campaign for girl's and women's empowerment.

Nowadays, Gucci Group is owned by the French-holding company Kering and has employed Marco Bizzarri as its CEO since January 2015. Very recently, Marco Bizzarri started to move from the aforementioned charitable activities to initiatives aimed at improving environment sustainability. For example, he announced that the Gucci brand will go fur-free next year and auction off its remaining animal fur items. The changes will come into force with the brand's spring–summer 2018 collection, Bizzarri said during a talk at the London College of Fashion, and reflect the firm's *"absolute commitment to making sustainability an intrinsic part of our business."* As part of the change, the fashion house will hold a charity auction for its remaining animal fur items and donate the proceeds to the animal rights organizations Humane Society International and LAV. Kitty Block, president of Humane Society

International, celebrated the luxury brand's move as a *"compassionate decision"*. Indeed, she underlined that *"Gucci going fur-free is a huge game-changer,"* and emphasized that *"for this Italian powerhouse to end the use of fur because of the cruelty involved will have a huge ripple effect throughout the world of fashion."*

In line with this announcement, Bizzarri engaged in another CE initiative aimed at strengthening Gucci's brand image as environmentally sustainable: the introduction of newly designed, completely recyclable luxury Gucci packaging. In particular, these newly designed bags, boxes, and tissue papers eschew the use of plastic laminate surfaces. Furthermore, the company began using cotton instead of polyester for its ribbon and garment bags. In fact, the luxury brand is cutting down on excess packaging altogether: Shoes will be packed in one flannel instead of two; gift boxes will only be given out when requested. In addition, Gucci is going to replace all of its mannequins with a new eco-friendly version. Designed by Frida Giannini and made in Italy, these new mannequins will be composed of shockproof polystyrene—a long-lasting and 100% recyclable raw material—and finished with water-based paints.

Beyond packaging alterations, Gucci is pursuing energy-saving initiatives in its retail stores. The firm aims to reach the following targets by the end of this year: a reduction of 35 tons of plastic waste; a reduction of 1,400 tons of paper consumption; a reduction of about 10,000 tons of CO_2 emissions, and a reduction of about 4 million liters of gas oil consumption.

In sum, Gucci's environmental orientation and waste-prevention initiatives represent CE-related activities that a top luxury brand can pursue. Meanwhile, the firm's ecological attention demonstrates a willingness to operate in a morally correct manner, and thus constitutes an ethical CSR activity.

6 Entrepreneurship and CE Initiatives Based on External (Legal and Philanthropic) CSR Dimensions

6.1 The Case of Stella McCartney

Born and raised in London, Stella McCartney started to work in fashion at the age of 15 with the French designer Christian Lacroix. After her graduation at London's Central Martins College of Art, in 1995, she decided to join the fashion house of Chloé in Paris. Two years later, she was selected to replace the departing head designer, Karl Lagerfeld. Her first collection in 1997 received enthusiastic reviews and laid the foundation for the common denominator of her style: sexy femininity and natural confidence.

In 2001, in partnership with Kering Group, Stella McCartney launched her own eponymous luxury lifestyle brand. She successfully introduced a series of creations that satisfied young women's desire to wear sexy, high-quality clothes, while also expressing their moods and independence in a stylish way. From the onset, Stella McCartney approached her business with a sustainability mindset. The brand is, in fact, internationally recognized for its entrepreneur's commitment to ethical values. For example, none of her clothing and accessory collections contains leather or fur; sandals are made with biodegradable soles; bags are lined with fabric made from recycled plastic bottles, and the eyewear collection uses natural and renewable components. Moreover, she refused to sell her fragrances in China because she would not allow her products to be tested on animals, which is a legal requirement in China. Stella McCartney considers her company to be responsible for the resources it uses and its impact on the environment; therefore, she is constantly exploring innovative ways to become more sustainable, from store practices to product manufacturing.

The brand's commitment to finding new ways to reduce its environmental impact, namely through recyclable and renewable materials, represents a valuable CE perspective. For instance, the brand is highly committed to exclusively using *regenerated* wool cashmere for her clothes. In fact, whereas a single cashmere sweater requires fiber from four goats, a manufacturer can make nearly five sweaters from the wool of one sheep. Cashmere has thus been traditionally considered a precious fiber, one that has experienced increasing demand from the luxury industry in recent decades. Exponential growth in the demand for cashmere raised serious environmental concern, especially in Mongolia, one of the main producers of cashmere. Since the 1990s, the goat population in Mongolia has increased fivefold; because goats graze avidly, their overpopulation has transformed once-rich grasslands into deserts. According to the United Nations Development Programme, 90% of Mongolia is fragile dry land that is under increasing threat of desertification.

Conscious of this alarming situation, the Stella McCartney company decided to replace all of its virgin cashmere knitwear with a regenerated cashmere yarn called Re.Verso™. Made in Italy, Re.Verso™ is a new integrated production process that is 100% traceable. The brand relies on select partners to acquire yarns, fabrics and knitwear, all of which are re-engineered into wool, cashmere and camel.

According to the EP&L (Environmental Profit & Loss account), cashmere made up only 0.13% of Stella McCartney's overall raw material usage in 2015, but still accounted for 25% of its total environmental impact. By solely using Re.Verso™ regenerated cashmere for the brand's 2016 needs, Stella McCartney estimated, the impact of its cashmere would fall to only 2%. Additionally, the brand is working with partners such as the Sustainable Fibre Alliance and the Wildlife Conservation Society to support on-the-ground work in Mongolia that can begin reversing the existing desertification.

Stella McCartney's attention to using regenerating cashmere is another clear example of a luxury entrepreneur brand pursuing a CE activity. At the same time, this activity can be interpreted as a legal CSR activity, since the cashmere regeneration was made possible through the Re.Verso™ certification.

6.2   The Case of Godiva

Godiva Chocolatier is a manufacturer of premium fine chocolates and related products. Founded in 1926 by Joseph Draps, the firm soon became a symbol of luxury in the department stores of Belgium. Campbell Soup Company acquired the full rights to the Godiva chocolatier company in 1974. In the USA, the brand maintained its luxury image as *"the Gucci of chocolate."* For the next three decades, Godiva remained focused on North America, from where it was controlling its overseas markets. In 2007, the firm was acquired by Yıldız Holding, owner of the Ülker Group and chaired by Murat Ülker, a Turkish industrialist and businessman internationally recognized for his attention to philanthropic initiatives. In 2009, for instance, Ülker established the Sabri Ulker Foundation with the goal of making a lasting contribution to public nutrition. In 2014, he contributed $24 million on behalf of the Ülker family to the Harvard T.H. Chan School of Public Health (HSPH) to establish the Sabri Ülker Center for Nutrient, Genetic, and Metabolic Research.

Under the Ülker, Group Godiva maintained its position, gaining a market share of 80% in the luxury chocolate segment (Deshpande and Çekin 2015). Nowadays, the brand owns and operates more than 600 retail boutiques and shops in the USA, Canada, Europe, and Asia, and is available via over 10,000 specialty retailers. In 2012, Godiva opened the Café Godiva in London's Harrod's Department Store: The 40-plus table venue features Godiva's chocolate beverages, pastries, and chocolates. The company also has a store in the Harrod's Food Hall.

Beyond improving Godiva's market coverage and revenues, Murat Ülker tries to imbue the Godiva brand with his humanitarian philosophy. He strives to invest in programs and practices that lead to lasting positive outcomes for Godiva consumers, partners and communities. One notable example is Godiva's membership to the World Cocoa Foundation (WCF), a leading nonprofit organization that promotes sustainability in the cocoa sector by providing cocoa farmers with the support they need to grow more quality cocoa and strengthen their communities. The WCF's work has led to increased productivity and profits for cocoa farmers, helping to ensure a sustainable supply of cocoa for generations to come. Godiva also participates in the Cocoa Horizons Foundation, which seeks to improve the livelihoods of cocoa farmers and their communities through the promotion of sustainable, entrepreneurial farming, improved productivity, and community development.

Another important sustainability project for Godiva is the "FEED Godiva Origins Collection". Started in 2012, this philanthropic activity involves reusing revenues in line with a CE perspective. In particular, Godiva works with the FEED company to produce and distribute a line of charitable, limited-edition chocolates that are typically sold during holiday seasons (e.g., Christmas or Mother's Day). These chocolates are inserted into a special gift bag that is sold at all Godiva boutiques, the revenues from which are used to help feed impoverished children, specifically in cocoa-sourcing regions.

This charitable initiative is a clear example of applying economic profit to a philanthropic cause—namely supporting children in developing countries. Grounded in Ülker's sustainable sensitivity, this CE initiative can be interpreted as a philanthropic CSR activity.

7 Conclusions: Managerial Implications, Limits and Future Research

The present chapter examined the CE initiatives undertaken by the entrepreneurs of four luxury companies: on the one hand Brunello Cucinelli S.p.A. and Gucci (characterized by CE activities related to internal CSR dimensions), on the other hand Stella McCartney and Godiva (characterized by CE activities related to external CSR dimensions). The most important conclusion that can be drawn from our analysis is that even in a prestige-based sector such as luxury, sustainability, and in particular CE-related actions, can be a fundamental recipe for success. In other words, luxury brands may base their entire branding strategy on sustainable values, thus anchoring their manufacturing uniqueness to innovative ideas that are consistent with the concepts of material recycling and reuse. Indeed, all four luxury brand cases presented in this chapter demonstrate that following the requirement of CE does not mean diluting luxury brands' quintessential prestige, heritage, and quality; rather, it means broadening the positive contribution that luxury brands are inherently able to make to the society in which operate.

In a recent publication (Amatulli et al. 2017a, b), the authors introduced the idea that luxury brands can in many cases be considered as *inherently* more sustainable than mass market brands, essentially because they have a higher potential to do good for employees, environment, economy, community in which they are located and society at large. Through the analysis presented in this chapter, we clarified this concept even more by discussing four cases of luxury companies that have been able to abide by the CE paradigm through the reuse of resources, thereby making a positive contribution to the environment, the employees and the society. In particular, we can summarize the cases presented saying that Cucinelli S.p.A. and Godiva are examples of *reuse* of economic resources generated in companies' market activities to mainly benefit such two relevant stakeholders as employees and producers' communities, while Gucci and Stella McCartney are examples of *reuse* of raw material to mainly benefit the environment. A common trait across all these cases is the key role played by the entrepreneur (often the founder of the company) in fostering the constant balance between the three Ps of the Triple Bottom Line (TBL) approach (Elkington 1994): People, Profit, Planet. Through the cases presented, it could be possible to add a fourth P, that of Place, since Godiva, in particular, is an example of how a luxury brand can do good also for the local communities of farmers.

From a theoretical point of view, this study opens new research opportunities related to sustainable luxury development. By exploring how CSR initiatives may be developed in light of both the CE approach and innovative entrepreneurship, our study strengthens the idea of CSR multidimensionality. Moreover, this research contributes to the literature on sustainable luxury by exploring how different CSR dimensions may play different roles in luxury consumption. Finally, this chapter advanced the literature on CE and entrepreneurship by connecting those concepts to both luxury and CSR.

From a managerial point of view, this study suggests luxury managers that sustainability should not only be considered as an important area to grant the company a positive status in the society, but it should be considered as quintessential to the development of their business and to the positive contributions they can make to different stakeholders. This is particularly relevant to understand in today's era in which all customer, including luxury ones, are increasingly sensitive to sustainability issues. Specifically, this chapter suggests that luxury brand manager can develop a thoughtful and well-planned CSR strategy by pursuing and monitoring activities (perhaps via internal progress reports) that balance the CE approach with the entrepreneurial initiatives that underlie sustainability. Managers may go the extra step of developing dashboards that assess their CSR efforts in order to maximize their commitment to sustainability.

Lastly, our research has some limitations that represent potentially interesting avenues for future studies. First, we focused on only four luxury brands, particularly in the fashion and food industries. Future studies may consider exploring the same research issue with more companies and/or different luxury sectors. Indeed, the shape of CE processes and influence of entrepreneurship may change based on the brands' industry and geography. Second, it would be useful to complement our explorative, qualitative approach with more structured research hypotheses and quantitative studies. Third, in aligning the luxury consumption dichotomy and the CSR dichotomy, we assumed that externalized (internalized) luxury consumers may be more attracted to external (internal) CSR dimensions. However, there have been no empirical demonstrations of this trend so far, which merits future investigations.

References

Achabou, M. A., & Dekhili, S. (2013). Luxury and sustainable development: Is there a match? *Journal of Business Research, 66*(10), 1896–1903.

Alvarado-Herrera, A., Bigne, E., Aldas-Manzano, J., & Curras-Perez, R. (2015). A scale for measuring consumer perceptions of corporate social responsibility following the sustainable development paradigm. *Journal of Business Ethics*, 1–20.

Amatulli, C., De Angelis, M., Costabile, M., & Guido, G. (2017a). *Sustainable luxury brands: Evidence from research and implications for managers*. Berlin: Springer.

Amatulli, C., De Angelis, M., Korschun, D., & Romani, S. (2017b). *Consumers' perceptions of luxury brands' CSR initiatives: An investigation of the role of status and conspicuous consumption*. Working paper.

Amatulli, C., & Guido, G. (2012). Externalised vs. internalised consumption of luxury goods: Propositions and implications for luxury retail marketing. *The International Review of Retail, Distribution and Consumer Research, 22*(2), 189–207.

Amatulli, C., Guido, G., & Nataraajan, R. (2015). Luxury purchasing among older consumers: Exploring inferences about cognitive age, status, and style motivations. *Journal of Business Research, 68*(9), 1945–1952.

Baumgartner, R. J. (2014). Managing corporate sustainability and CSR: A conceptual framework combining values, strategies and instruments contributing to sustainable development. *Corporate Social Responsibility and Environmental Management, 21*(5), 258–271.

Bhattacharya, C. B., & Sen, S. (2004). Doing better at doing good: When, why, and how consumers respond to corporate social initiatives. *California Management Review, 47*(1), 9–24.

Bielak, D., Bonini, S. M., & Oppenheim, J. M. (2007). CEOs on strategy and social issues. *McKinsey Quarterly, 10*(1), 8–12.

Brown, T. J., & Dacin, P. A. (1997). The company and the product: Corporate associations and consumer product responses. *The Journal of Marketing*, 68–84.

Carroll, A. B. (1979). A three-dimensional conceptual model of corporate performance. *Academy of Management Review, 4*(4), 497–505.

Carroll, A. B. (1991). The pyramid of corporate social responsibility: Toward the moral management of organizational stakeholders. *Business Horizons, 34*(4), 39–48.

D'Aprile, G., & Mannarini, T. (2012). Corporate social responsibility: A psychosocial multidimensional construct. *Journal of Global Responsibility, 3*(1), 48–65.

Davies, I. A., Lee, Z., & Ahonkhai, I. (2012). Do consumers care about ethical-luxury? *Journal of Business Ethics, 106*(1), 37–51.

Deloitte Development. (2011). Sustainability strategy 2.0—Next-generation driver of innovation. http://www.deloitte.com/assets/Dcom-UnitedStates/Local%20Assets/Documents/us_scc_SustainabilityStrategy_061311.pdf.

D'Anolfo, M., Amatulli, C., De Angelis, M., & Pino, G. (2017). Luxury, sustainability, and corporate social responsibility: Insights from fashion luxury case studies and consumers' perceptions. In *Sustainable management of luxury* (pp. 427–448). Singapore: Springer.

Deshpande, R., & Çekin, E. (2015). Rebranding Godiva: The Yıldız Strategy. *Harvard Business Case*.

Elkington, J. (1994). Towards the sustainable corporation: Win-win-win business strategies for sustainable development. *California Management Review, 36*(2), 90–100.

European Commission. (2015). Closing the loop-An EU action plan for the Circular Economy. *Communication from the Commission to the European Parliament, the Council, the European Economic and Social Committee and the Committee of the Regions COM, 614*(2), 2015.

Geissdoerfer, M., Savaget, P., Bocken, N. M., & Hultink, E. J. (2017). The circular economy–A new sustainability paradigm? *Journal of Cleaner Production, 143*, 757–768.

Gershoff, A. D., & Frels, J. K. (2015). What makes it green? The role of centrality of green attributes in evaluations of the greenness of products. *Journal of Marketing, 79*(1), 97–110.

Han, Y. J., Nunes, J. C., & Drèze, X. (2010). Signaling status with luxury goods: The role of brand prominence. *Journal of Marketing, 74*(4), 15–30.

Kapferer, J. N. (2010). All that glitters is not green: the challenge of sustainable luxury. *European Business Review*, 40–45.

Langenwater, G. A. (2009). Planet first. *Industrial Management, 51*(4), 10–13.

Lochard, C., & Murat, A. (2011). *Luxe et développement durable: La nouvelle alliance*. Editions Eyrolles.

Luo, X., & Du, S. (2015). Exploring the relationship between corporate social responsibility and firm innovation. *Marketing Letters, 26*(4), 703–714.

Maignan, I., & Ferrell, O. C. (2001). Corporate citizenship as a marketing instrument—Concepts, evidence and research directions. *European Journal of Marketing, 35*(3/4), 457–484.

Moffett, M. H., & Ramaswamy, K. (2003). Fashion faux pas: Gucci & LVMH. *Thunderbird International Business Review, 45*(2), 225–239.

Mohr, L. A., & Webb, D. J. (2005). The effects of corporate social responsibility and price on consumer responses. *Journal of Consumer Affairs, 39*(1), 121–147.

Mohr, L. A., Webb, D. J., & Harris, K. E. (2001). Do consumers expect companies to be socially responsible? The impact of corporate social responsibility on buying behavior. *Journal of Consumer Affairs, 35*(1), 45–72.

Nidumolu, R., Prahalad, C. K., & Rangaswami, M. R. (2009). Why sustainability is now the key driver of innovation. *Harvard Business Review, 87*(9), 57–64.

Nueno, J. L., & Quelch, J. A. (1998). The mass marketing of luxury. *Business Horizons, 41*(6), 61–68.

Pino, G., Amatulli, C., De Angelis, M., & Peluso, A. M. (2016). The influence of corporate social responsibility on consumers' attitudes and intentions toward genetically modified foods: evidence from Italy. *Journal of Cleaner Production, 112,* 2861–2869.

Shane, S., & Venkataraman, S. (2000). The promise of entrepreneurship as a field of research. *Academy of Management Review, 25*(1), 217–226.

Yuan, Z., Bi, J., & Moriguichi, Y. (2006). The circular economy: A new development strategy in China. *Journal of Industrial Ecology, 10*(1–2), 4–8.

Carmela Donato is a Post-Doc student of Marketing at LUISS University, Italy, and has been lecturer at RUG University in the Netherlands. She presented her research in the main international marketing conferences as ACR, EMAC, AMSWC. Moreover she has published in major international refereed journals as Psychology & Marketing. Her research topics are mainly focused in sustainable luxury and green consumption.

Cesare Amatulli is Assistant Professor of Marketing at the University of Bari, Italy. He has been a Visiting Professor at LUISS University, Italy, and Visiting Researcher at the Ross School of Business, University of Michigan, US, and at University of Hertfordshire, UK. He has published articles in major international referred journals, such as European Journal of Marketing, Journal of Business Research, Journal of Business Ethics and Psychology & Marketing. He is also author of the book titled Sustainable Luxury Brands.

Matteo De Angelis is Associate Professor of Marketing at LUISS University, Italy, and has been a Visiting Scholar at Kellogg School of Management and Visiting Professor at the University of Wisconsin, USA. His articles have been published in several leading marketing journals such as Journal of Marketing Research, Journal of Consumer Research, Journal of the Academy of Marketing Science, Journal of Business Ethics, International Journal of Research in Marketing, Journal of Business Research and Psychology & Marketing. He is also author of the Book titled Sustainable Luxury Brands.

Challenging Current Fashion Business Models: Entrepreneurship Through Access-Based Consumption in the Second-Hand Luxury Garment Sector Within a Circular Economy

Shuang Hu, Claudia E. Henninger, Rosy Boardman and Daniella Ryding

Abstract The purpose of this chapter is to investigate drivers of (non)participation in access-based consumption and the underpinning motives of becoming (or not) a micro-entrepreneur within the circular economy. Peer-to-peer platforms and drivers of (non)participation within the context of the UK's second-hand luxury market are currently under-researched. This chapter is exploratory in nature and utilises a qualitative research approach. This study conducts semi-structured interviews with consumers from varied demographical backgrounds to gain an insight into consumers' perceptions of access-based consumption and sustainability. Findings identified drivers of (non)participation and the emergences of a (potentially) new micro-entrepreneur. It is further explored whether this would be a feasible business model for the future with consumers actively buying into the access-based concept. Although findings cannot be generalised, the data provides thinking points for future research and investigates an economically significant context. Gaining an insight into this newly emerging trend could help retailers to capitalise on disruptive innovations and change consumer perceptions of partaking in access-based consumption. Thus far, drivers of (non)participation in the context of the UK's second-hand luxury industry remain under-researched, and the economic significance of the sector indicates the necessity of this research.

S. Hu (✉) · C. E. Henninger · R. Boardman · D. Ryding
School of Materials, University of Manchester, Oxford Road, Manchester M13 9PL, UK
e-mail: shuang.hu@postgrad.manchester.ac.uk

C. E. Henninger
e-mail: claudia.henninger@manchester.ac.uk

R. Boardman
e-mail: rosy.boardman@manchester.ac.uk

D. Ryding
e-mail: daniella.ryding@manchester.ac.uk

© Springer Nature Singapore Pte Ltd. 2019
M. A. Gardetti and S. S. Muthu (eds.), *Sustainable Luxury*,
Environmental Footprints and Eco-design of Products and Processes,
https://doi.org/10.1007/978-981-13-0623-5_3

39

Keywords Sustainability · Access-based consumption · Second-hand luxury fashion · Business model · Renting · Circular economy · Micro-entrepreneur

1 Introduction

This chapter provides new insights into the motivational drivers of (non)participation in access-based consumption, a topic currently under-researched (Park and Armstrong 2017). Access-based consumption is part of the circular economy and fosters a relatively new business model in the fashion industry, by providing access to, in this case, second-hand luxury items, for a limited amount of time in exchange for a set fee. Within the circular economy, sharing practices are more and more commonplace, whereby peer-to-peer (P2P) facilitation becomes key. This aspect of P2P sharing is further investigated in this chapter, by analysing consumer motivations to participate (renting), as well as actively partake (renting out) in these business models. The latter aspect links to entrepreneurialism and individuals taking advantage of their idle resources (e.g., unused garments and accessories). The emergence of a new micro-entrepreneur within the circular economy has been observed, yet not fully investigated in terms of motivational drivers (Bucher et al. 2016), a further gap addressed in this chapter. Thus, we pose the following questions:

1. What are the associations and perceptions of luxury fashion and access-based consumption models in the context of the UK's second-hand luxury fashion industry?
2. To what extent are these P2P and business-to-consumer models associated with sustainability?
3. What are drivers of these business models and how can they be further enhanced? An analysis of the emergence of a new type of entrepreneurs.

Sustainability is currently a buzzword within the fashion industry and has gained increased importance over the past decades (e.g. Joy et al. 2012; Henninger et al. 2016). The luxury fashion industry has been under pressure to incorporate more sustainability practices, especially after the "Slaves of Luxury" RAI3 programme in 2007, which highlighted social sustainability issues in the supply chain (cited in Gardetti and Torres 2017), or more recently animal welfare issues, such as utilising exotic leather in garment and accessories production and manufacturing processes (Butler 2017). Although it could be argued that, aside from obvious environmental and social issues, luxury fashion is more "sustainable" because it provides garments or accessories that cater towards a niche market that are often rare and unique, the emergence of the fast fashion phenomenon, which sees fortnightly garment collection turnarounds, has fostered a massification of luxury products in the form of entry-level items, thereby opening it up to the mass-market. This implies that luxury fashion, in a similar vain to fast fashion, is fostering hyper consumerism, which will not be sustainable in the future (Thomas 2007; Wang and Griskevicius 2014).

Although a rather bleak backdrop for the industry, these aspects have led to an entrepreneurial spirit and developed an opportunity for new business models to emerge within a circular economy. The circular economy challenges current system thinking and moves away from a linear to a re-looped, or cradle-to-cradle approach within the supply chain, in which materials are used and re-used more frequently before being discarded (McDonough and Braungart 2002; Niinimäki 2013; Ghisellini et al. 2016). One entrepreneurial business model within the circular economy is that of access-based consumption, which allows the facilitation of, in this case, renting second-hand luxury items in either a P2P or business-to-consumer (B2C) environment. Access-based consumption implies that, rather than transferring ownership of a product, consumers can make temporary use of them, which extends their usefulness and ensures the maximum amount of uses of a product prior to being "discarded". It is a market-mediated transaction to make use of goods/services without seeing a transfer of ownership (Bardhi and Eckhardt 2012), by simply paying a standard fee (Schaefers et al. 2016). Access-based models allow sharing and pooling of resources/products/services (Botsman and Rogers 2010; Belk 2014). In this manner, renting implies that garments are utilised on a more frequent basis and shared by multiple people, thus ensuring that products get used and re-used more often than would be the case if a single individual owned the product. Although access-based consumption models are not new per se (Airbnb and Uber are prime examples), they are under-researched and newly emerging within the UK second-hand luxury industry.

Internet technologies have further fostered the facilitation of access-based consumption and enabled new emerging business models to rapidly become global phenomena (Botsman and Rogers 2010; Stephany 2015). As indicated, access-based consumption models are commonplace in the tourism (Airbnb, CouchSurfing), transportation (Uber), and more recently, the fashion industry (Rent the Runway, Girls meet Dress) (Perlacia et al. 2016). Rude (2015) advocates that the social community, economic growth, sustainable development and practical improvement drive individuals to engage in collaborative consumption, which provides an opportunity for the fashion industry to enhance sustainability by decreasing the production of clothing and minimising the amount of waste (Perlacia et al. 2016). Clifford (2011) indicates that fashion retailers are starting to capitalise on access-based consumption models and, more generally, think about embracing the circular economy by providing renting pop-up events, swap shops or second-hand sales of their own fashion lines. We focus on the consumer's ability to temporarily access second-hand luxury goods such as fashion clothing, bags and jewellery, without the transferring of ownership of these items.

The chapter provides a brief overview of the economic significance of the second-hand luxury fashion industry, before reviewing the importance of access-based consumption and what is currently known about drivers and barriers in the renting market. This chapter is exploratory in nature and presents initial findings of our data collection, critically discussing these before providing implications and ideas for future research.

2   Literature Review

2.1   *Sustainability and Second-Hand Luxury Fashion*

The most commonly used definition of sustainability is meeting the needs of the current generation, without putting barriers in place to fulfil future generations' needs (e.g. Fletcher 2008). Sustainability has been a part of the fashion repertoire since its inception and is therefore not a new concept (Gardetti and Torres 2013). Henninger et al. (2016) point out that sustainability and fashion could imply an idiosyncrasy, as fashion suggests something that always changes, is trend-led, and this, may be discarded of rather quickly. Contrarily, sustainability implies a long-term perspective, which further incorporates aspects of the social, environmental, and economic facet (Gardetti and Torres 2013). Due to documentaries, such as "Slaves of luxury" (RAI3 cited in Gardetti and Torres 2013) or the dispatches programme "Undercover: Britain's cheap clothes" (Channel 4 2017), fashion organisations are now under pressure to make sustainability a top priority within their business strategies (Walker 2006; Henninger et al. 2016, 2017a).

The term *luxury* implies an indulgence, as it literally translates to "extras of life" (Danziger 2005). Li et al. (2012) state that luxury consists of desirable objects that provide pleasure and belong to the top category of prestigious brands (Vigneron and Johnson 2004). Luxury brands are classified as high quality, expensive, and consist of copyrighted products/services that have high levels of symbolic and emotional value that are seen to be rare, exclusive, prestigious, and authentic (Tynan et al. 2010). Yet, what is luxury is subjective and may vary depending on who is questioned (Phau and Prendergast 2000). Similar to the luxury fashion sector, the second-hand luxury market, and the second-hand market in more general terms, has seen an increase in sales over recent years, with the global second-hand apparel market being worth £14 billion and forecasted to grow by approximately 11% annually to become a £25 billion industry by 2021 (Kestenbaum 2017).

Second-hand fashion implies that a garment and/or accessory has previously been worn and gains a new lifespan by entering the second-hand industry, which links to the aspect of sustainability in that "waste" materials are re-used and their life extended (e.g. Cervellon et al. 2012; Henninger et al. 2016). We include vintage fashion in the second-hand market, although it is acknowledged that not all vintage items are necessarily second-hand (Cervellon et al. 2012). A key point of differentiation between vintage and second-hand is the era of when they (garments/accessories) were produced; vintage garments stem from the 1920s to 1980s and in some cases may have never been worn, whilst second-hand can incorporate any era (Mortara and Ironico 2011; Cervellon et al. 2012). Although second-hand and luxury apparel has been investigated in past research (Isla 2013; Turunen and Leipämaa-Leskinen 2015; Turunen 2017), only the forthcoming edited book *Vintage luxury fashion: Exploring the rise of secondhand trading* (Ryding et al. forthcoming) explores aspects of access-based consumption and second-hand luxury fashion in more detail. Yet, our overall understanding of the implications of what access-based consumption has on the future and what the underpinning drivers for participation and non-participation are remain limited (Bucher et al. 2016; Park and Armstrong 2017; Ryding et al. forthcoming).

Although the luxury fashion market continues to grow, reflected in high profit margins, it also faces a variety of challenges, such as fakes, insufficient natural resources and a lack of disposable income on the consumer side (Karaosman et al. 2017). The latter aspect could explain why access-based consumption models have gained popularity, as consumers are still able to indulge in luxury fashion consumption without having to pay the full retail price for the product. At the same time, conscious shoppers are able to still have "new" luxury products, without partaking in hyper consumerism, by simply renting an item. A question that arises and is addressed in this chapter is: To what extent are people who own luxury products willing to share their garments/accessories with others for a fee? In line with necessity being the mother of invention, a new type of micro-entrepreneur has emerged, one who capitalises on idle resources—but what are their underling motivations for partaking in this new business adventure?

2.2 Motivational Drivers, Sustainability and Access-Based Consumption

A key question addressed in this chapter centres on (non)participation in P2P renting platforms for second-hand luxury fashion. Hence, it is vital to explore the motivational drivers of (1) consuming luxury items and whether this translates into second-hand luxury consumption and (2) partaking in access-based consumption, which also links to entrepreneurialism and capitalising on unused resources.

Motivational drivers for luxury fashion consumption have predominantly focused on the "first-hand" rather than the second-hand market (e.g. Wang and Griskevicius 2014; Henninger et al. 2017b), thereby identifying conspicuousness, snob effect, bandwagon effect and sociality as key aspects. On the other hand, Cervellon et al. (2012) provide an insight into the second-hand and vintage market, which sees eco-consciousness, frugality, the need for status and uniqueness, fashion involvement and nostalgia emerging as dominant factors, swaying consumers to purchase these items. Although the context of past studies differs, with some focusing on first-hand and others on second-hand markets (but not necessarily luxury), an overlap can be observed in terms of the underpinning drivers as to why these items may be consumed, which are summarised in Table 1. Access-based consumption is a newly emerging phenomenon in the fashion industry, and thus far, only a view studies have indicated potential motivational drivers, such as monetary/economic benefits, moral responsibility, social-hedonic aspects and environmental considerations (Table 1). These, however, need to be explored further (Bucher et al. 2016; Park and Armstrong 2017).

Table 1 Motivational drivers for luxury fashion and access-based consumption (adapted from Henninger et al. 2017a, b)

Motivation	Meaning	Reference
Luxury goods purchase—motivational drivers		
Conspicuousness (need for status)	• Acquiring wealth to enhance social standing	Veblen (1889), Wang and Griskevicius (2014)
Snob effect (need for uniqueness)	• Portraying self-image and social identity through unique, individual, and exclusive items	Bian and Forsythe (2012), Kastanakis and Balabanis (2014)
Bandwagon	• Need to showcase social belonging to a specific group	Tsai (2005)
Sociality	• Broadcasting social standing in society through materialism	Chen and Kim (2013), Wang and Griskevicius (2014)
Nostalgia proneness	• Emotional attachment to garment	Cervellon et al. (2012), Armstrong and Page (2015)
Fashion involvement	• The extent to which fashion is personally relevant	Cervellon et al. (2012)
Frugality	• Monetary aspects	Cervellon et al. (2012)
Eco-consciousness	• Feeling a responsibility to protect environment	Cervellon et al. (2012)
Access-based consumption—motivational drivers		
Monetary/economic	• Making use of idle capacities to gain additional income • Gaining new items for less money	Belk (2014), Bellotti et al. (2015), Bucher et al. (2016)
Moral responsibility	• Need to give back to society and sharing out wealth	Armstrong and Page (2015), Park and Armstrong (2017)
Social-hedonic	• Social aspect of partaking in access-based consumption	Akbar et al. (2016), Armstrong and Page (2015)
Environmental	• Conscious consumption and actively seeking to reduce environmental impact	Armstrong and Page (2015)

Although Table 1 is not a systematic review of current literature, it provides a snapshot of what is currently known about luxury and second-hand/vintage consumption and participation in access-based consumption and, more specifically, renting garments. It becomes apparent that economic drivers, social standing and sociality, as well as environmental consciousness, are a common thread within the literature. Whether these motivational drivers are the same for consumers and micro-entrepreneurs and how these translate into the context of the second-hand luxury fashion industry are further investigated in this chapter.

Table 2 Summary of data collection (authors' own)

Participant ID	Gender	Age	Occupation
I1	Male	20	University student
I2	Female	32	Ph.D. student
I3	Female	26	Masters student
I4	Male	28	Owner of commercial company
I5	Female	18	A level student
I6	Male	48	Owner of business company
I7	Male	23	Masters student
I8	Female	25	Marketing director
I9	Male	30	Civil engineer
I10	Female	36	Energy consultant
I11	Male	24	Engineering consultant
I12	Female	21	Masters student
I13	Female	25	Masters student
I14	Female	25	Building services engineer
I15	Male	26	Ph.D. student
I16	Male	27	Shop supervisor

3 Methodology

This interpretivist research is exploratory in nature and utilises a qualitative research approach to explore drivers of (non)participation in access-based consumption and motives to participate and partake in renting second-hand luxury clothing. As part of an on-going research project, this study conducts in-depth semi-structured interviews with 16 consumers from varied demographical backgrounds to gain an insight into consumers' perceptions of access-based consumption, sustainability and entrepreneurship. Participants for this research were recruited using convenience and snowball sampling, meeting the following criteria: (1) aged between 18 and 55 years, as they represent consumers that have considerable purchasing power and/or disposable income to consume luxury goods (Drapers 2013); (2) UK based; (3) either consume luxury goods and/or have rented luxury goods previously. The third criteria provide a better understanding of the current and potential market, which allows us to explore participation and non-participation in renting second-hand luxury garments, with interviewees being split equally between males and females. Table 2 provides a summary of our participants.

Interview questions were guided by our literature review and focused on aspects of sustainability and its meaning and motivational drivers of (non)participation in renting second-hand luxury garments. Moreover, we investigated consumer perceptions and explored the underpinning entrepreneurial spirit of those participants who have previously rented out their own belongings. All interviews were conducted face-to-face and lasted on average 30 min.

Qualitative research is inductive, and it emphasises the meaningfulness of the research (Bodgan et al. 2015). The rich qualitative data sets from this study were coded and re-coded multiple times by the researchers and followed Easterby-Smith et al.'s (2012) grounded theory approach based on a seven step guide: familiarisation, reflection, open-coding, conceptualisation, focused re-coding, linking and re-evaluation. In order to guarantee continuity, coherence and clarity, the researchers first analysed the data independently, focusing on phrases and words most commonly mentioned by interviewees and explored within their natural boundaries. The open-coding process allowed for patterns to emerge naturally.

A limitation of this research could be the interpretivist nature of the study undertaken and the limited number of participants. Yet, findings provide an interesting insight into the circular economy, sustainability and entrepreneurialism, which leads to valuable thinking points that are discussed in the following section.

4 Findings and Discussion

4.1 Luxury Fashion and Sustainability

As indicated in the literature review, "luxury" is a subjective term and can have different meanings for different people (Phau and Prendergast 2000). In order to have a common ground for discussion, it was thus seen as important to gain an insight into what "luxury" meant to our research participants. A majority of interviewees indicate that *luxury* is about high quality, fashionability, and durability of products, stating: "*luxury goods are of really good quality and stylish […] when I purchase luxury items I feel good as this is a high quality product, which I will probably use in the long-term*" (I1). I4 further highlights "*when I own or use luxury goods, the brand image, as well as the product features, make me feel… admired […] and it makes me happy as I am proud of myself that I can afford such an expensive item*", which links to aspects of conspicuous consumption (Wang and Griskevicius 2014).

Data indicates that the interviewees did not necessarily have a preference of luxury brands (e.g. LV, Hermès, Gucci), as long as they are well-known, have a good reputation, portray a respected brand image and provide high-quality products and services (I4, I8, I13, I14, I15, I16). Although some may be inclined to purchase slightly less well-known brands, I16 states that there are risks associated with it, as it is only if a brand has a "*good reputation… I know what standard of products to expect*". It becomes apparent that participants believe luxury products are of good quality and can be worn for a longer period of time, which indirectly suggests that this way of consumption might be more sustainable, as there is a long-term commitment involved (Gardetti and Torres 2013).

As a result of the luxury fashion industry being scrutinised in the press recently and negative connotations emerging, we explored whether aspects of sustainability influence UK consumers' purchasing decisions. Similar to results found by Fletcher (2008), we found that interviewees associate environmentally friendly products, economic benefits (e.g. cost saving) and fair working environments with sustainability. Indeed, the data indicates that it (sustainability) plays a vital role for these participants within their daily routines and business lives, with I4 insisting "*for my business, sustainability means a healthy cash flow... stock control, exclusive market potential, and efficient operation... For my customers, sustainability means I can offer great services and achieve their needs in the long term*". Although interviewees state that they are trying to make conscious choices, this may not always be translated into their luxury purchase decision, as "*luxury purchases are usually a one-off*" (I12) and a reward for hard work undertaken. Several of our interviewees felt strongly about the material that is used, especially in the luxury industry, and boycott purchasing anything that contains fur or exotic leather. "*I do not want to purchase products that are made of fur or leather from a rare or protected species*" (I9), as, although he enjoys owning garments and accessories from luxury brands, he draws a line between what is luxury and what is morally wrong. Similarly I2, I3, I13 and I15 have consciously started to avoid products from endangered species, especially after learning more about sustainability and fashion production at university.

Charitable engagement and being viewed as a "good corporate citizen" further attracts I15, who "*prefer(s) products that are recycled... and have a reusable value... some brands or products engage with charities... that makes me feel good*". This was further mentioned by I6 and I16. Interestingly, charitable causes only seemed to have been mentioned by our male participants. Although exploring this aspect further is beyond the scope of this chapter, it provides an avenue for future research.

Unlike prior research that suggests that sustainability does not play a key role in the decision-making process (e.g. Henninger et al. 2017a, b) our research found a shift in attitude whereby interviewees felt passionate about environmental aspects and animal welfare. To reiterate this further, having had his own business for multiple years, I4 has become more aware of issues across industries, highlighting that it is the final stage—the discarding of products—that primarily captures his attention. "*I will also consider the potential risk for the environment when I have no use for a product and need to throw it out... I will consider the potential group who might need it or whether it can be recycled*", which implies that I4 either donates items or actively seeks to keep them out of landfill where possible.

An interesting finding is that, although participants state that they are making conscious decisions, they also believe that luxury fashion and sustainability are two concepts that are not compatible (I3, I6, I9, I12, I15). I15 goes as far as to say that luxury companies, who utilise animal skin are "*evil(sig)... (they) destroy the eco-system and the environment*". On the other hand, I14 believes that "*the cost of luxury goods... their cost on the environment may be low as by their nature they are produced in small quantities*" and although leather and fur may be used in the production process, the overall impact may be negligible (I4, I11). Two very strong groupings emerge in the data set: one who could be described as eco-warriors and protectors of animal welfare and one who seeks to find excuses to make themselves feel better about indulging in luxury products. This chapter does not seek to provide a judgement of whether indulging in luxury products is right or wrong, but rather tries to portray two sides to an issue, which could also underpin why participants might either be already renting or are inclined to rent and participate in access-based consumption. This aspect is further explored in the next section.

4.2   Second-Hand Luxury Fashion and Access-Based Consumption

The previous section provided an overview of luxury fashion in general terms, whereas this section explores underpinning motives for participating in the second-hand movement and engaging with access-based consumption. From the findings reported earlier, it became apparent that there are some concerns, in terms of the materials used within luxury fashion production, which causes a divide within our data set. As such, we explored whether second-hand luxury purchases could provide an alternative to overcoming these challenges and to what extent access-based consumption will be a viable option for our participants.

The majority of participants believed that second-hand luxury fashion has had a positive impact and is something that will be of benefit in the future. "*I think it's a good idea as someone can buy something which is cheaper… and it still has the brand name […] it's not as expensive as brand new. However, the quality must be the same as it was brand new or I would not buy it*" (I1). This suggests that as long as the quality of a product adheres to the same standards as a new product purchased from the same brand, consumers will be inclined to purchase it. Interviewees further highlight that at times you cannot actually differentiate between first and second-hand products:

> there are a lot of second-hand fashion shops in London and Apps such as Depop and Vinted… some second-hand retailers have exceptional cleaning processes as I cannot tell whether it's new or second-hand… some organisations offer cleaning, maintenance, and repair services for luxury fashion. (I2)

This statement vocalises concerns raised by other participants also, in that they like the idea of second-hand fashion, however, were afraid of issues such as hygiene (I3, I10, I11). Moreover, interviewees felt uneasy not knowing whether they were buying something new or second-hand, as there could be hidden damage to the product or it could be counterfeit, an aspect that was also raised by previous research (Cervellon et al. 2010), indicating that even experts at times struggle to identify authentic pieces. This has various implications for luxury brands in terms of their brand image and reputation. If a consumer unknowingly purchases a second-hand item that breaks after only a few uses, the consumer may not trust the brand. Similarly, if they were led to believe they bought an authentic vintage item that turned out to be an exceptional reproduction of an original piece, and thus a fake, this can lead to a damaged brand image.

Aside from these challenges, interviewees, especially I2 and I10, felt passionate about second-hand items, as they are more environmentally friendly and provide an opportunity to extend the useful life of a product to at least one further owner (Guiot and Roux 2010). The data further indicates that several of our participants felt positive about access-based consumption and, especially, the renting of luxury products, as they saw a link between being sustainable, but still gain the benefit of showcasing their favourite luxury brands, which supports past research (Gleim and Lawson 2014). *"Renting is a good idea... the company can make good profits on this as they can continually use and redistribute their products... and the consumer could also save money"* (I1), which suggests a win–win situation. Interviewee I9 further insists that

> renting something means that you can use it when you need it and return it when you do not need it... others can do the same... renting makes full use of the garments... it is wasteful for everybody to own these items of clothing and it is not environmentally sustainable.

Our findings concur with previous research (Belk 2014; Armstrong and Page 2015; Bellotti et al. 2015; Bucher et al. 2016) stating that access-based consumption is linked to monetary benefits, as luxury products can be utilised by paying a small fee, as well as environmental aspects, by gaining the most use out of a garment. In addition, it seems predominantly male participants that are interested in sustainable aspects and who see benefits of renting and other alternative modes of consumption. The male interviewees seem to be responsive to trying new things and are more conscious of the impact their purchases have on the environment. Why this may be the case is beyond the scope of this chapter, but again provides an avenue for further research.

4.3 Motivations of Access-Based Consumption and Entrepreneurial Spirit

Data suggests that both economic and environmental aspects are key motivational drivers to participating in access-based consumption, as consumers can showcase products without having to spend a lot of money. Those interviewees that do not have any previous renting experiences could only see themselves renting luxury handbags or expensive occasion wear. Similar to Lastovicka et al. (1999) and Eastman et al. (1999), we found that price-conscious consumers were inclined to engage in renting, as this allows interviewees to show off their social status and portray a version of their ideal self that they aspire to be in the future. I12 further highlights that *"usually if something is really expensive, I rent it before I buy it"* (I12). This indicates that access-based consumption could be motivated by risk aversion and seeking to postpone a purchase decision, in order to ensure that they have made the right choice (Matzler et al. 2007; Schwartz 2015). This is a novel finding, as it has not previously been suggested that avoiding risk could be a reason to participate in access-based consumption.

Unlike in other industries, such as transportation (Uber) or the tourism and hospitality sector (Airbnb), several interviewees highlighted that they would not trust renting an item from a stranger: "*I prefer to rent products from third-party agents as the third-party will identify and detect... if the products are genuine and not damaged*" (I2). The interviewee continues: "*renting services can offer legal protection for damaged luxury items... or provide rules and regulations for when goods are damaged... otherwise there's a lot of risk involved... both for renting and rent out*". The risks associated with renting and renting out a luxury garment are perceived to be higher than those in other industries, as interviewees highlight that these luxury items represent parts of their personality. If they are damaged, fake or negatively marked in any other way, this has a direct impact on themselves and their social standing (I2, I3).

Those interviewees renting out their luxury items predominantly highlighted that it had monetary benefits, which concurs with past research (Belk 2014; Bellotti et al. 2015; Bucher et al. 2016). I4 further highlights that, for him, renting out his underutilised luxury items is a creative idea to generate stories and rank items. The more people that use his items, the more stories these products can tell and the more emotional value they have. Items that he rents out on a frequent basis are thus charged at a higher fee, as the value of the item increases in emotional terms. To explain, if a bag was used at an interview and the person renting the bag got the job, others may believe it is a lucky bag that will enable them to achieve their aspirations. Thus, emotional capital is a key driver of participating in access-based consumption. I6 also highlights that he feels that renting out his luxury products provides those who are less wealthy access to luxury items, making them feel special. However, he states that he only rents out items that he no longer wants to use, as he is afraid that items might get damaged and would not want this to happen to an item that he is still attached to.

5   Conclusion

This chapter was set out to investigate drivers of (non)participation in access-based consumption, thereby exploring three key questions: (1) perceptions of access-based models in the second-hand luxury industry, (2) associations of access-based models with sustainability and (3) drivers to (non)participating in access-based consumption and underpinning entrepreneurial spirits.

Our data indicates that there is generally a positive attitude towards access-based consumption and the idea of renting garments and accessories. Interviewees felt it was a good compromise of still being able to showcase their social status and belonging, whilst at the same time have an opportunity to be more sustainable. Interviewees had fewer reservations about renting items that are produced with controversial materials, as these were classified as vintage and thus had already been in the market for quite a while. Making use of these products was seen as a way of extending their useful life and not wasting natural resources used in the production process.

A key contribution presented in this chapter is the characteristic differences which exist between access-based consumption in the luxury second-hand market and other industries. In the second-hand luxury market, it is evident from the data presented that consumers want reassurance that the garments and accessories are rented out by official agents, rather than individual private entities. As such, business models such as Airbnb, where consumers can also be suppliers, could potentially fail, as there is a lack of trust and security associated with this approach in the fashion industry. A further novel difference to the access-based consumption model in the second-hand luxury fashion industry is the way in which people use renting to eliminate associated risks, before buying an expensive item. Renting, in this sense, has helped consumers with their decision-making by acting as a precursor and trial run for a big purchase. Finally, the research found clear gender differences in attitudes towards sustainability as a driver for access-based consumption. In particular, men were much more motivated by sustainability benefits to use access-based consumption models in the second-hand luxury fashion industry, which provides an interesting avenue for further research.

This research is based on findings from a limited amount of in-depth semi-structured interviews and is qualitative in nature; thus, there is potential to extend this study in the future and explore its reliability. With access-based consumption remaining an under-researched phenomenon in the fashion context, future research is needed to further explore the similarities and differences across industrial boundaries. Clear differences emerge between P2P and B2C rental platforms in the second-hand luxury fashion industry, when compared with the tourism and transportation industry. Further investigations are necessary in order to gain a deeper understanding of why disruptive innovations have not yet fully capitalised on the entrepreneurial spirit in the second-hand luxury fashion industry.

Our research further alluded to differences between male and female participants with regard to whether charitable causes are a part of the sustainability agenda, having only been mentioned by male participants. Due to the scope of the chapter, this finding could not be explored further, yet would be interesting to investigate. It further seemed that men were highly responsive to trying out new things, such as access-based consumption, as well as being highly conscious of the environmental impact their decisions have on the natural environment and society at large. It would be interesting to explore this aspect further and why this may be the case.

References

Akbar, P., Mai, R., & Hoffmann, S. (2016). When do materialistic consumers join commercial sharing systems. *Journal of Business Research, 69,* 4215–4224.

Armstrong, A., & Page, N. (2015). Creativity and constraint: Leadership and management in the UK creative industries. *Creative Skillset.* Retrieved January 26, 2018 from https://creativeskillset.org/assets/0001/5933/Creativity_and_constraint_leadership_and_management_in_UK_2015.pdf.

Bardhi, F., & Eckhardt, G. M. (2012). Access-based consumption: The case of car sharing. *Journal of Consumer Research, 39,* 881–898.

Belk, R. (2014). You are what you can access: Sharing and collaborative consumption online. *Journal of Business Research, 67*(8), 1595–1600.

Bellotti, V., Ambard, A., Turner, D., Gossmann, C., Demkova, K., & Carroll, J. M. (2015). *A muddle of models of motivation for using peer-to-peer economy systems*, CHI 2015, April 18–23, 2015, Seoul: Republic of Korea.

Bian, Q., & Forsythe, S. (2012). Purchase intention for luxury brands: A cross cultural comparison. *Journal of Business Research, 65*(1), 1443–1451.

Bodgan, R., Taylor, S. J., & DeVault, M. (2015). *Introduction to qualitative research methods: A guidebook and resource.* Hoboken, New Jersey: John Wiley & Sons Inc.

Botsman, R., & Rogers, R. (2010). *What's mine is yours—How collaborative consumption is changing the way we live.* London: Harper Collins Publishers.

Bucher, E., Fiesler, C., & Lutz, C. (2016). What's mine is yours (for a nominal fee)—Exploring the spectrum of utilitarian to altruistic motives for internet-mediated sharing. *Computer in Human Behaviour, 62,* 316–326.

Butler, S. (2017) Gucci owner gest teeth into snakeskin market with python farm. *The Guardian.* Retrieved October 26, 2017 from https://www.theguardian.com/business/2017/jan/25/gucci-snakeskin-python-farm-kering-saint-laurent-and-alexander-mcqueen.

Cervellon, M.-C., Carey, L., & Harms, T. (2012). Something old, something used: Determinants of women's purchase of vintage fashion vs second-hand fashion. *International Journal of Retail & Distribution Management, 40*(12), 956–974.

Channel 4. (2017). Undercover: Britain's cheap clothes: Channel 4 Dispatches. *Channel 4.* Retrieved September 18 2017 from http://www.channel4.com/info/press/news/undercover-britains-cheap-clothes-channel-4-dispatches.

Chen, J., & Kim, S. (2013). A comparison of Chinese consumers' intentions to purchase luxury fashion brands for self-use and for gifts. *Journal of International Consumer Marketing, 25*(1), 29–44.

Clifford, E. (2011). *Consumer attitudes towards luxury brands.* UK, November 2011. Mintel, London.

Danziger, P. (2005). *Let them eat cake: Marketing luxury to the masses—As well as the classes.* New York: Dearborn Trade Publisher.

Drapers. (2013). Drapers luxury report 2013. *Drapers.* Retrieved January 26, 2018 from https://www.drapersonline.com/Journals/2015/07/09/h/k/h/2013-Nov-16-Luxury-Report.pdf.

Easterby-Smith, M., Thorpe, R., & Jackson, P. R. (2012). *Management research.* London: Sage.

Eastman, J. K., Goldsmith, R. E., & Flynn, L. R. (1999). Status consumption in consumer behaviour: Scale development and validation. *Journal of Marketing Theory and Practice, 7*(3), 41–52.

Fletcher, K. (2008). *Sustainable fashion and textiles: Design journeys. Sustainable fashion and textiles: Design journeys.* Earth Scan: London.

Gardetti, M. A., & Torres, A. L. (Eds.). (2013). *Sustainability in fashion and textiles: Values, design, production and consumption.* Sheffield, UK: Greenleaf Publishing.

Ghisellini, P., Cialani, C., & Ulgiati, S. (2016). A review on circular economy: The expected transition to a balanced interplay of environmental and economic systems. *Journal of Cleaner Production, 114,* 11–32.

Gleim, M., & Lawson, S. J. (2014). Spanning the gap: An examination of the factors leading to the green gap. *Journal of Consumer Marketing, 31*(6/7), 503–514.

Guiot, D., & Roux, D. (2010). A second-hand shoppers' motivation scale: Antecedents, consequences, and implications for retailers. *Journal of Retailing, 86*(4), 383–399.

Henninger, C. E., Alevizou, P. J., Goworek, H., & Ryding, D. (2017a). *Sustainability in fashion—A cradle to upcycle approach.* Heidelberg: Springer.

Henninger, C. E., Alevizou, P. J., & Oates, C. J. (2016). What is sustainable fashion? *Journal of Fashion Marketing and Management, 20*(4), 400–416.

Henninger, C. E., Alevizou, P. J., Tan, J., Huang, Q., & Ryding, D. (2017b). Consumption strategies and motivations of Chinese consumers—The case of UK sustainable luxury fashion. *Journal of Fashion Marketing & Management, 21*(3), 419–434.

Isla, V. L. (2013). Investigating second-hand fashion trade and consumption in the Philippines: Expanding existing discourses. *Journal of Consumer Culture, 13*(3), 221–240.

Joy, A., Sherry, J. F., Jr., Venkatesh, A., Wang, J., & Chan, R. (2012). Fast fashion, sustainability, and the ethical appeal of luxury brands. *Fashion Theory, 16*(3), 273–295.

Karaosman, H., Brun, A., & Morales-Alonso, G. (2017). Vogue or vague: Sustainability performance appraisal in luxury fashion supply chains. In *Sustainable management of luxury* (pp. 301–330).

Kastanakis, M. N., & Balabanis, G. (2014). Explaining variations in conspicuous luxury consumption: an individual differences' perspective. *Journal of Business Research, 67*(10), 2147–2154.

Kestenbaum, R. (2017). Fashion retailers have to adapt to deal with secondhand clothes sold online. *Forbes.* Retrieved January 26, 2018 from https://www.forbes.com/sites/richardkestenbaum/2017/04/11/fashion-retailers-have-to-adapt-to-deal-with-secondhand-clothes-sold-online/#476d48901a7f.

Lastovicka, J. L., Bettencourt, L. A., Hughner, R. S., & Kuntze, R. J. (1999). Lifestyle of the tight and frugal: Theory and measurement. *Journal of Consumer Research, 26*(1), 85–98.

Li, G., Li, G., & Kambele, Z. (2012). Luxury fashion brand consumers in China: Perceived value, fashion lifestyle, and willingness to pay. *Journal of Business Research, 65*(10), 1516–1522.

Matzler, K., Waiguny, M., & Füller, J. (2007). Spoiled for choice: Consumer confusion in internet-based mass customization. *Innovative Marketing Journal, 3*(3), 7–18.

McDonough, W., & Braungart, M. (2002). *Cradle to cradle: Remaking the way we do things.* New York, USA: North Point Press.

Niinimäki, K. (2013). *Sustainable fashion: New approaches.* Helsinki, Finland: Aalto ARTS Books.

Park, H., & Armstrong, C. M. J. (2017). Collaborative apparel consumption in the digital sharing economy: An agenda for academic inquiry. *International Journal of Consumer Studies, 41,* 465–474.

Perlacia, A. S., Duml, V., & Saebi, T. (2016). *Collaborative consumption: Live fashion, don't own it* (Doctoral dissertation, Norwegian School of Economics).

Phau, I., & Prendergast, G. (2000). Consuming luxury brands: The relevance of the 'rarity principle'. *Journal of Brand Management, 8*(2), 122–138.

Rude, L. (2015). 4 keys to a successful sharing economy business model. *Text100.* Retrieved January 26, 2018 from https://www.text100.com/2015/05/11/sucessful-sharing-economy-business-model/.

Ryding, D., Henninger, C. E., & Blazquez Cano, M. (forthcoming). *Vintage luxury fashion: Exploring the rise of secondhand clothing trade.* Palgrave Advances in Luxury Series. London: Springer.

Schaefers, T., Wittkowski, K., Benoit, S., & Ferraro, R. (2016). Contagious effects of customer misbehavior in access-based services. *Journal of Service Research, 19*(1), 3–21.

Schwartz, A. (2015). Regulating for rationality. *Stanford Law Review, 67*(6), 1373–1410.

Stephany, A. (2015). *The business of sharing: Making it in the new sharing economy. The business of sharing: Making it in the new sharing economy.* Basingstoke: Palgrave.

Thomas, D. (2007). *Deluxe: How luxury lost its luster.* New York: Columbia University Press.

Tsai, S. (2005). Impact of personal orientation on luxury-brand purchase value: An international investigation. *International Journal of Market Research, 47*(4), 427–452.

Turunen, L. L. M. (2017). *Interpretations of luxury: Exploring the consumer perspective.* Basingstoke, UK: Palgrave.

Turunen, L. L. M., & Leipämaa-Leskinen, H. (2015). Pre-loved luxury: Identifying the meanings of second-hand luxury possessions. *Journal of Product & Brand Management, 24*(1), 57–65.

Tynan, C., McKechnie, S., & Chhuon, C. (2010). Co-creating value for luxury brands. *Journal of Business Research, 63*(11), 1156–1163.

Veblen, T. (1889). *The theory of the leisure class.* Boston, MA: Houghton Mifflin.

Vigneron, F., & Johnson, L. W. (2004). Measuring perceptions of brand luxury. *Journal of Brand Management, 11*(6), 484–506.

Walker, S. (2006). *Sustainable by design: Explorations in theory and practice.* London: Earthscan.

Wang, Y., & Griskevicius, V. (2014). Conspicuous consumption, relationships, and rivals: Women's luxury products as signals to other women. *Journal of Consumer Research, 40*(5), 834–854.

Shuang Hu is a Ph.D. student in Textile Design, Fashion and Business at the School of Materials, the University of Manchester. Her research focuses on sustainability, access-based consumption and service quality in the context of the luxury fashion industry. She presented her recent work at Transitioning to Sustainability Conference (CE2S) hosted by Centre for Business in Society, Coventry University, UK.

Dr. Claudia E. Henninger is Lecturer in Fashion Marketing Management at the School of Materials, The University of Manchester. Her research centres on corporate marketing, sustainability, and green consumer typologies in the context of the fashion industry. She has published in leading journals, contributed to edited books, and presented her work at various national and international conferences. She is Deputy Chair of the Academy of Marketing SIG Sustainability.

Dr. Rosy Boardman is Lecturer in Fashion Business at the School of Materials, The University of Manchester. Rosy's research primarily focuses on digital strategy and innovation in the retail industry. In particular, her research specialises in e-commerce, digital marketing, multichannel/omnichannel retailing, m-commerce, consumer behaviour, sustainability and the circular economy, utilising eye-tracking technology and qualitative research methods. Rosy regularly attends academic conferences presenting her work and has published peer-reviewed academic journal papers.

Dr. Daniella Ryding is Senior Lecturer in Fashion Marketing at the School of Materials, The University of Manchester, UK. As an active researcher with over 40 published articles and several text book contributions, her main research interests focus on consumer behaviour within an international fashion retailing context. She takes a proactive interest in sustainability and the circular economy, examining further the ecological and ethical principles which drive business strategy.

Sadhu—On the Pathway of Luxury Sustainable Circular Value Model

Sheetal Jain and Sita Mishra

Abstract The concepts of 'luxury' and 'sustainability' are antithetical to each other. However, a paradigm shift is presently witnessed in luxury domain. Lately, sustainability is swiftly becoming a critical issue for both luxury brands as well as society as a whole. This case study focuses on the company—'Natweave Textile Studio.' It is a textile company founded by Indian textile designer Subhabrata Sadhu in 2009, with a yearning to conserve the rich heritage of rarest and finest cashmere by using the traditional skills of native Kashmiri artisans. The company specializes in producing high-end and exclusive Pashmina scarves and shawls with focus on entirely pure, handmade, and natural production process. Sadhu sources finest Pashmina fibers from Pashmina goats reared in its natural habitat in Changthang plateau in the Kashmir region. He strongly believes in protecting and preserving the rare art form and providing a platform to the Kashmiri craftsmen—custodians of this ancient skill, to showcase their culture to the world. He collaborates with Kashmiri weavers to create contemporary products and remodel ethnic weaves into timeless luxury items and works hard to combine traditional techniques with modern designs to build sustainable luxury products. This study aims to develop a luxury sustainable circular value (LSCV) model that integrates the values of four stakeholders—entrepreneur, organization, customers, and society. LSCV model is used as a tool to examine how 'Natweave Textile Studio' contributes toward creating sustainable circular value and thus adds to the sustainable development of the company and society.

Keywords Circular economy · Entrepreneurship · India · Kashmir · Luxury
Pashmina · Sustainability · Value chain

S. Jain (✉)
Luxe Analytics, IILM, Lodhi Road, New Delhi, India
e-mail: bardiaconsulting@gmail.com

S. Mishra
Marketing, Institute of Management Technology, Ghaziabad, UP, India
e-mail: sitamish@gmail.com

© Springer Nature Singapore Pte Ltd. 2019
M. A. Gardetti and S. S. Muthu (eds.), *Sustainable Luxury*,
Environmental Footprints and Eco-design of Products and Processes,
https://doi.org/10.1007/978-981-13-0623-5_4

1 Introduction

Sustainable circular luxury invigorates the revival of rare art forms and provides unique platform for artisans and craftsmen to conserve their local culture and integrate social and environmental issues in the entire value chain. Entrepreneurs play a crucial role in developing sustainable circular luxury by bringing transmutation for betterment of people and planet. They work on principles of circular economy and create restorative economic model, to produce timeless luxury items which are inherited through generations.

This case presents one such company 'Natweave Textile Studio' which is committed to conserve the 2000 years old legacy of Pashmina and sustainable methods of production. Starting from sourcing to dyeing to spinning and weaving no automated and chemical-driven processes are used to ensure that fabric holds its natural softness and warmth and it lasts forever. It helps in understanding the drivers of sustainable circular luxury business by integrating the values of four stakeholders—entrepreneur, organization, customers, and society.

2 Methodology

This case study was conducted in two phases. In the first phase, bibliographic compilation on the given subject was done. In the second phase, data about case company was collected through trade media as well as through semi-structured interviews with founder and the artisans.

3 Luxury and Sustainability: Close Relationship

The concept of luxury is based on high quality, exclusivity, heritage, timelessness, superior craftsmanship, and rarity (Jain et al. 2015; Kapferer 2010; Brun et al. 2008; Djelic and Ainamo 1999). Kapferer (2013) argued that luxury possesses features of sustainability and referred it 'the business of lasting worth and permanence.' Sustainability is considered as a prerequisite for any luxury brand because luxury buyers pay premium price not just to get superior quality but also they expect luxury fashion retailers to contribute toward social and environmental responsibility. Ducrot-Lochard and Murat (2011) posited that luxury industry is moving toward sustainability as luxury buyers are looking for high-quality products which provide no harm to environment. Kim and Ko (2012) contemplated that luxury brands can establish their association with consumers not only on the basis of their brand name and the intrinsic quality or exclusivity of their products but also by taking into consideration the humanitarian and the environmental values. Cervellon and Shammas (2013) contended that luxury can be considered as complementary to sustainability in case consumers perceive the brand as 'making luxury' in terms of craftsmanship, unique resources, and anchorage in its origins. Steinhart et al. (2013) demonstrated that an environmental perspective of brands positively enhances consumers' perceptions of both utilitarian and luxury products. Consumers may go beyond this by either recompensing or reprimanding companies that focus or overlook the importance of social and environmental excellence (Grail Research 2010). Sustainable development provides an opportunity for enhancing brand image and differentiation of luxury brands as consumers are showing their interests and awareness toward social and environmental issues (Kim et al. 2012). Realizing the importance of this concept, many luxury firms are redefining their supply chain processes, attaining efficiency, nurturing innovation, and enhancing focus on their brand (Kapferer and Bastien 2012).

Literature on luxury describes it as something that is 'more than necessary' (Bearden and Etzel 1982), having an 'intensely individual component' (Berthon et al. 2009), 'superfluous' (De Barnier et al. 2006), and associated with 'dream' (Seringhaus 2002). Further, the prominence on attributes such as heritage, timelessness, durability, and production in smaller batch sizes with slow production cycle (Henninger et al. 2017) gives strong argument for luxury to be associated with sustainability in a multidimensional value approach. Yet, luxury industry has been criticized of ignoring their social and environmental responsibilities (Charles 2010; Nair 2008; Siegle 2009). It is evident from past studies that luxury industry trails behind other industries when it comes to responsibility toward sustainability (Bendell and Kleanthous 2007).

4 Luxury and Sustainability: Converse Relationship

The term 'Luxury' for long is linked with extravagance, superfluous, non-essential, image, and status, which combine to make it desirable for reasons other than function (Wiedmann et al. 2009; Vigneron and Johnson 2004; Dubois et al. 2005). Basic values of sustainability—humanitarianism, restraint, and temperance are in complete contrast with luxury's inherent values of hedonism, rarity, superfluity, and aestheticism (Carrier and Luetchford 2012). Consequently, literature advocates negative perception about sustainable luxury goods and envisages these as less desired in comparison to regular non-sustainable luxury products. Achabou and Dekhili (2013) examined consumers' inclination toward recycled materials in French luxury clothing purchase and found that integrating recycled materials in this category of goods influenced adversely their preferences. In addition, a brand's environmental commitment was the least imperative criterion for purchasing luxury goods while quality, price, and brand reputation were the most preferred. Likewise, Davies et al. (2012) established that consumers consider minimal impact of sustainability and ethics during decision-making process of luxury products in comparison to commodity consumption. However, luxury goods were perceived as relatively more sustainable when assessed against a context of high-street consumption. Torelli et al. (2012) indicated in their study that consumers experience a sense of disfluency to a responsible luxury brand which communicates about CSR in comparison to a brand which provides no CSR information. They found that in prior case, the brand is not perceived appropriately and gets lesser evaluation. However, the effect may differ with the relative conspicuousness of the branding strategy followed by a luxury brand.

Looking into the concepts of 'luxury' and 'sustainability,' it is clear that they are antithetical to each other. However, a paradigm shift is presently witnessed in luxury domain (Bendell and Kleanthous 2007; Davies et al. 2012; Kendall 2010). As the number of well-educated, socially responsible, affluent, global elite is rapidly rising, the concept of sustainability is becoming top priority for luxury brands (Cvijanovich 2011). Large number of luxury fashion retailers including Gucci, Prada, and Armani, to name a few, are responding to the increased consumer demand for making the supply chain transparent and sustainable (Joy et al. 2012; Henninger et al. 2016). Luxury brands are trying to reposition themselves as 'the caretaker of mother earth' to create a favorable brand image (Doval et al. 2013).

5   Sustainability and Circular Economy

The broadly acknowledged definition of sustainable development is 'development which meets the needs of the present without compromising the ability of future generations to meet their own needs' (WCED 1987, p. 43). Its foundation is based upon the notion that resources are limited and have to be managed to sustain next generations. It has been long time that industries have been looking for counseling in implementing strategies for sustainable development. The focus on sustainability requires an overhaul change from magnitude of goods produced to the quality of human knowledge, ingenuity, and self-realization as measures of development. Additionally, the emphasis on quality of life, human cohesion, and ecological sensibility should be more as these are the leading tools in the transition process (Möller 2006). It is in this context that a new approach to sustainability, the 'Circular Economy' is appearing as a possible solution to companies of all sizes to establish socially responsible business. Cisco's President, Chris Dedicoat described Circular Economy as 'a blueprint for a new sustainable economy, one that has innovation and efficiency at its heart and addresses the business challenges presented by continued economic unpredictability, exponential population growth and our escalating demand for the world's natural resources' (Ellen MacArthur Foundation 2013).

Feng et al. (2007) envisaged the circular economy as an approach of economic development which relies on ecological circulation of natural materials and needs compliance with environmental laws and consumption of natural resources to accomplish economic development. The Circular Economy has been the most recent effort to integrate the three pillars of sustainable development—social, economic, and environmental (Murray et al. 2017). Serious attempts are being made to move from traditional linear economic model (take-make-dispose) to circular economy through embedding the 3R principle (Reduce, Reuse, and Recycle) into production and consumption process (Zhu and Qui 2007; Heck 2006). However, Fletcher (2008) described that reuse and recycling strategies may not be very useful in certain product categories (viz. clothes) whose life cycle is enormously short. Nevertheless, by making consumers cognizant about resources and effort on making clothes, companies can make consumers understand the reverence of products and reduce their discard (Hope 2015).

Circular economy makes organizations less dependent on primary commodities, offers opportunities for new revenues, and creates value. The value creation in circular economy in comparison with linear product design and materials usage is based on 'power of the inner circle,' 'power of circling longer,' 'power of cascaded use,' and 'power of pure circles' (Ellen MacArthur Foundation 2013). Circular economy focuses on high efficiency, greater energy savings, and low greenhouse gas emission. Therefore, circular economy can be redefined as a new concept of closed-loop economy, value, procurement, production, restorative, redesigning, and consumption, resulting in sustainable development of economy, planet, and people (Wu 2005; Shen 2007). The circular economy focuses on augmenting systems rather than constituents. Moreover, it stresses not only on customary ideas and strategies of sustainability but also on diminishing the harm to natural ecosystems and reestablishment of the environment (Pitt 2011). Largely, circular economy concept has generated interests from a variety of stakeholders, created knowledge and skills toward a more sustainable society (Genovese et al. 2017).

6   Sustainable Circular Luxury and Entrepreneurship

Sustainable luxury encourages the revival of rare art forms and provides unique platform for artisans and craftsmen to preserve their local culture and integrate social and environmental issues in production process (Gardetti and Torres 2013). Luxury creates value from objective rarity which may be because of usage of rare materials or to a rare craftsmanship. The safeguard of exceptional craftsmanship contributes to spawning luxury's rarity. Timeless luxury items favor endurance and reduce waste and thus have an impact on environmental sustainability. Entrepreneurship plays a crucial role in promoting sustainable circular luxury. These entrepreneurs are transformational leaders as they work for betterment of people and planet (Gardetti and Torres 2013). They work on principles of circular economy and create restorative economic model (Ellen MacArthur Foundation 2013), to produce timeless luxury items which are inherited through generations.

Luxury industry is also influenced by the sharing economy concept wherein new technologies, economic downturn, and concern for a more responsible way of consuming play vital role. Few luxury players are working toward new business models and ensure a more efficient management of scarce resources, modifying the process of design production by reducing waste in the production process and using materials that have a lower impact on the environment. Many entrepreneurs in luxury industry have realized the importance of sustainability and thus are remodeling their traditional businesses by incorporating concepts of reuse, share, rent, or recycle which are essence of circular economy. The trendsetters of sustainable luxury are also using technology to connect with crafts workers and customers and deliver genuine luxury product based on 3Rs principles of circular economy. Luxury players like Alan Frampton of Cred, Michael Stober of Landgut Stober Business Hotels, Isobel Davies of Izzy Lane and many more pioneers are showing great interest in adopting sustainable practices (Gardetti and Girón 2017). The Kering Group in order to ensure transparency presents an interactive statement on its Web site by illustrating the various steps in production and environmental categories where it is making an impact.

7 Creating Luxury Sustainable Circular Value

Bansal and Roth (2000) suggested values, economic opportunities, legislation, and stakeholder pressures as four crucial motives for corporations focusing on sustaining the natural environment. The conceptions and definitions of term 'value' have been defined in varied ways in different academic disciplines and are frequently inconsistent (Rohan 2000). Rokeach (1973) defines 'a value is an enduring belief that a specific mode of conduct or end state of existence is personally or socially preferable to an opposite or converse mode of conduct or end state of existence' (p. 5). Meglino and Ravlin (1998) summarized values literature to illustrate that the evaluation of decisions and subsequent behavior is associated with individual values. Values are crucial for an organization's ethical system to become sustainability-oriented (Stead and Stead 2009). They are key predictors of pro-environmental behavior (Bansal 2003; Karp 1996). Elkington (1999) argues that an organization must understand that creation of social and environmental value is equally important as creation of business value. There is requirement for systematic change in all societal aspects, including values, consuming patterns, technological, and business innovations, in order to create sustainability and circularity (Ellen MacArthur Foundation 2013).

According to Joy et al. (2012), 'Concomitant respect for artisans and environment' promotes stronger pro-sustainability values among young luxury fashion consumers. The transition toward a circular economy requires a systematic change, in which organizations have to take the effects of their own business and the direct effects on its suppliers, customers, and other stakeholders into account, and vice versa. The business literature, which focuses on circular economy's business opportunities, is almost void. A lot of research is done on how value can be created through optimized utilization of materials nonetheless how intangible value is generated from circular economy is less researched area.

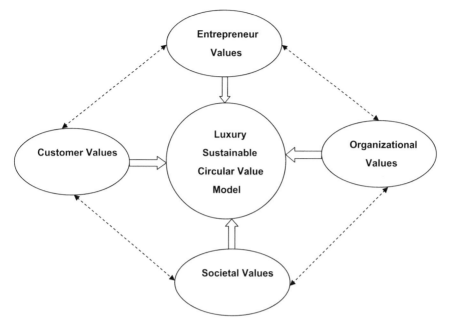

Fig. 1 Luxury sustainable circular value model (LSCV). *Source* Designed by authors

8 Luxury Sustainable Circular Value Model (LSCV)

This study aims to understand the drivers of sustainable circular luxury business by developing a comprehensive LSCV model that integrates the values of four stakeholders—entrepreneur, customer, organization, and society (Fig. 1). The following section will discuss how each of these values (entrepreneur values, customer values, organizational values, and societal values) drives luxury companies to work through 'LSCV lens.'

8.1 Entrepreneur Values

According to Shane and Venkataraman (2000), '…entrepreneurship is concerned with the discovery and exploitation of profitable opportunities' and 'involves the nexus of two phenomena: the presence of lucrative opportunities and the presence of enterprising individuals' (pp. 217–218). Entrepreneurs augmented interest regarding social and environmental issues has resulted in the emergence of new terminology in literature such as environmental entrepreneurs (Schaper 2002) or ecopreneurs (Isaak 1998) and social entrepreneurs (Dees 2001). The concept of sustainability entrepreneur has evolved nascent (Crals and Vereeck 2004) which is gaining attention. Crals and Vereeck (2004) defined sustainable entrepreneurship as the 'continuing commitment by businesses to behave ethically and contribute to economic development while improving the quality of life for the workforce, their families, the local, and global community as well as future generations.' Schick et al. (2002) revealed the influence of entrepreneur's knowledge, perception, and personal values in considering sustainability-related issues while exploiting new business opportunities.

Rokeach (1973) classified the 36 values equally into two groups and named them as 'terminal values' (which reveal the idealized end goals of an individual) and as instrumental values (which help an individual to achieve their end goals). These values have been accepted as unique values that entrepreneurs hold (Fagenson 1993) and have been steadily related with concerns about others' well-being (Agle et al. 1999; Hood 2003). European Multistakeholder Forum on CSR (2004) advocated the motive to incorporate sustainability into a small- and medium-sized enterprise (SME) is usually based on the values and personal beliefs of the founder(s). However, their ambition to achieve seems to be associated to a need for personal fulfillment with regard to intellectual and professional goals. On attaining such goals, the entrepreneur realizes a sense of self-respect (Chapman 2000) which may be more significant for them than making money (Corman et al. 1988).

Entrepreneurs' values entrench into the organization's policies and processes which may influence employees for long period, even after the founders' departure (Baron and Shane 2007). Entrepreneur values are crucial while formulating a firm's strategy (Fagenson 1993) and are manifested in its priorities, resulting in organizational process and outcomes (Berson et al. 2008). Many studies (Nonis and Swift 2001; Hemingway 2005) have explored importance of personal values and how these values influence social entrepreneurship and ethical decision making. Krishnan (2001) studied the association between transformational leadership with values that extend to collective welfare, equity, and moral values. On similar lines, Waldman et al. (2004) indicated that transformational leadership is valuable for execution of corporate social responsibility. Various entrepreneurs in luxury domain have been found to develop an inclusive supply chain with economically weak section of Latin America with an objective to provide them livelihood and promote their local culture by producing simultaneously environmentally sustainable products (Gardetti and Torres 2013). Figure 2 depicts entrepreneurial values that guide luxury companies to evaluate business opportunities through 'LSCV lens.'

8.2 Customer Values

Earlier, the desire of 'buying to impress others' was considered as primary motive for purchasing luxury brands (Wiedmann et al. 2007). Ramchandani and Coste-Maniere (2012) described consumption of sustainable luxury products in public as pro-environmental behavior and philanthropy. Though, they posited that this behavior might be true in countries like India and China where conspicuousness, imitative

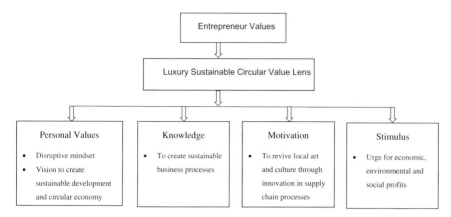

Fig. 2 Entrepreneur values to evaluate business opportunities through 'LSCV lens'. *Source* Designed by authors

behavior and word of mouth impacts consumption of luxury industry. However, in recent times, luxury brands have redefined their business models due to consumers' shift from 'conspicuous value' to 'conscientious value' (Cvijanovich 2011). Luxury brands are mostly used to signal consumer's self-concept and individuality (Belk 1988); therefore, luxury buyers 'want the brands they use to reflect their concerns and aspirations for a better world' (Bendell and Kleanthous 2007, p. 5). A luxury brand affiliated with the values of sustainability is viewed as a value enabler rather than big-headed. This association creates a sense of belongingness and cognizance that involves the conscious luxury consumers (Kale and Öztürk 2016).

Spangenberg (2001) considered that generally consumers do not get many opportunities to influence environmental issues, but clothing is one category where individuals can impact especially by creating demand for sustainable products. Burns (2010) specified that product categories which are linked with consumers' self-construction and identity are constantly assessed on both aesthetic and social grounds. Consumers prefer products with values that they want to connect to themselves and which have emblematic meanings that are connected to identity construction (Wang and Wallendorf 2006).

Moral and ethical issues can enhance opinion and self-perception and they constitute a progressively critical factor in the psychological satisfaction afforded by luxury goods (Olorenshaw 2011). Cervellon and Shammas (2013) comprehended sustainable luxury values in three categories—sociocultural values (conspicuousness, belonging, and national identity), ego-centered values (guilt-free pleasures, health and youthfulness, hedonism, durable quality), and eco-centered values (doing good, not doing harm). Consumers believe that it is the responsibility of luxury companies to address moral issues related to luxury products and provide 'convincing answers to questions of environmental and social responsibility' (Bendell and Kleanthous 2007, p. 8). In future, luxury consumers will be recognized as one who has the

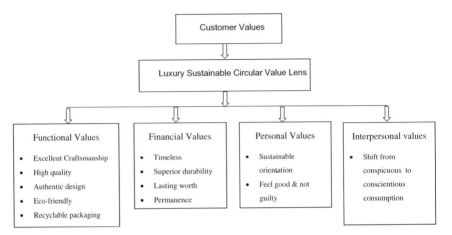

Fig. 3 Customer values that drive luxury companies to work through 'LSCV lens'. *Source* Designed by authors

resources as well as desire to care for societal and environmental benefit (Bendell and Kleanthous 2007).

Luxury sustainability customer values can be further classified as financial, functional, personal, and interpersonal values (Hennigs et al. 2013) (Fig. 3). Following is the brief description of each of these values that drive luxury companies to work through 'LSCV lens':

Financial Value: Consumers seek to achieve financial value through 'luxury products that will last, maintaining the brand's heritage into the future (Bendell and Kleanthous 2007, p. 29). Cooper (2005) describes durability and longevity as an essential characteristic for sustainable products but consumers generally link durability with high quality and not with environmental impact.

Functional Value: Kapferer and Bastien (2012) considered luxury to be linked with individual's egoistic pleasure yet sustainability commands rationality and focus on the functional values. The desire for high-quality, excellent craftsmanship, authentic design, and eco-friendly items drives sustainability (Kapferer 2010). Luxury brands associated and ensuing sustainable practices reverberate their values of endurance and longevity. These values stand for a longer period and help the company to position it on core values of craftsmanship (Kale and Öztürk 2016). Performance of the product has an influence on circular economy thinking, around 80% of sustainability-driven consumer segment 'would buy more sustainable products, if they performed comparable to and/or better than their usual brand' (Bemporad et al. 2012, p. 43).

Personal Value: There is a shift to from 'self-orientation' to 'sustainable-orientation' among luxury consumers (Bendell and Kleanthous 2007). They want to feel good and not guilty, when they are purchasing a certain luxury brand (Kendall 2010). They are seeking for the value beyond the product itself through investment in health and well-being (Hashmi 2017).

Interpersonal Value: There is gradual shift from focus on 'conspicuous value' to 'conscientious value' among the luxury consumers (Cvijanovich 2011). In addition, millennials want to display their social and environmental concerns behind luxury purchase decisions to the society through various social media platforms.

8.3 Organizational Values

The vision of sustainability and circularity, which is a must-have arsenal for future businesses, provides essential direction to the members of organization with respect to organizational priorities, technological advancements, resource allocation and business model development that creates value for the organization (Hart and Milstein 2003). An organization can create sustainable business value if it addresses following challenges: minimize waste from current operations and prevents pollution; acquire or develop more sustainable technologies and skill sets; engage in interaction with external stakeholders, with respect to both present offerings as well as profitable new solutions to future problems related to people and planet (Hart 2007). Bansal (2003) highlighted that an organization's reactions to environmental issues are influenced by the level of individual concern and by the values entrenched within the organization.

Cervellon and Shammas (2013) envisaged assimilation of corporate social responsibility and corporate environmental responsibility concepts which influence the strategic perspective, mission, and vision of many companies operating in luxury industry. Luxury companies are striving to reduce social and ecological problems through the use of sustainable technologies (Hart 2005, 2007) and pursuing sustainable operational practices. They are using environment-friendly raw materials, like organic cotton and natural dyes (Grail Research 2010); for instance, the leather of a Dior handbag is attained from Italian biofarms (Kapferer 2010). Many organizations are also using recyclable packaging to ensure sustainable circular luxury processes. Louis Vuitton with creation of LIFE (LVMH Initiatives for the Environment) in 2012 has commenced a comprehensive program including their production, procurement, and retail operations and beginning right from product design. The company transports 60% of its leather goods from France to Japan via ship, to avoid pollution and works toward saving energy by 30% at all new stores by implementing a new lighting concept (Grail Research 2010).

Sustainability gives a lot of focus on cultures and local traditions and provides attention to artisans, the savoir-faire of ancient traditions. Sustainable luxury is about nurturing craftsmanship, the handmade arts, and the local art and craft (Kale and Öztürk 2016). A number of luxury brands are now involved in supporting and contributing toward numerous social and environmental initiatives viz. Louis Vitton's social responsibility is based on four principles: workplace well-being and quality working conditions, developing talent and savoir-faire, preventing discrimination, and supporting local communities. Beard (2008) indicated that changing approach toward sustainable strategies viz. fair trade manufacturing may influence success of the firm. Consequently, companies like Kering focus on sustainable practices

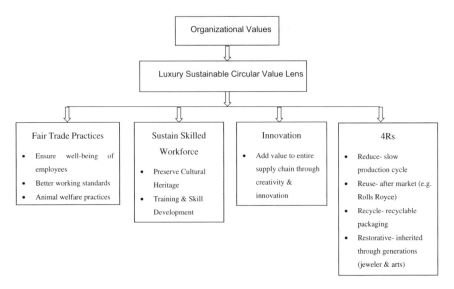

Fig. 4 Organizational values that drive luxury companies to run business processes through 'LSCV lens'. *Source* Designed by authors

such as reduction of carbon dioxide emissions, waste, and water; procurement of raw materials; elimination of hazardous chemicals and materials; and paper and packaging. Valente (2012) explained that 'what is sustained is a result of a complex interactive and idiosyncratic process where firms and their stakeholders build cognitive complexity within a network system in a way that creates synergistic value creation' (p. 586). Further, he pointed out that, for a 'sustain-centric ambition,' organizations in value networks have to collaborate in such way that it includes all relevant organizations and embraces all related systems (inclusion), understands all causes and effects of these systems in interrelationships (interconnectedness) and any position of privilege by 'a fair distribution of resources, opportunities, basic needs, and property rights' (equity). Figure 4 depicts the organizational values that drive luxury companies to run business processes through 'LSCV lens.'

8.4   Societal Values

Few studies (Bendell and Kleanthous 2007; Kleanthous 2011) highlighted that luxury brands are liable of shrinking social inequalities due to which their reputation could be declined. Consequently, consumers in all social classes are increasingly concerned about social and environmental issues. There has been paradigm shift in luxury realm as consumers are increasingly inclined toward sustainable orientation; not only in context of Western luxury markets but also among Asian, Latin American, and Eastern European markets (Bendell and Kleanthous 2007). Luxury brands have

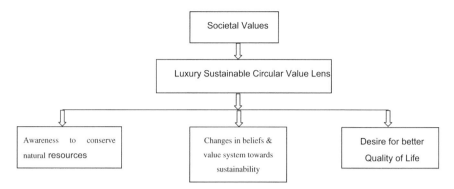

Fig. 5 Societal values that drive luxury companies to develop new business models through 'LSCV lens'. *Source* Designed by authors

the potential and raw material to spur innovation and progress toward well-being. Refocusing aspirations from connection with a product to association with a purpose results luxury consumption from a conspicuous effort into an intellectual endeavor.

Chapple and Moon (2005) analyzed the CSR reports of 50 firms in seven Asian countries and found variation. They emphasized that this dissimilarity in cross-country corporate sustainability cannot be solely described on the basis of a country's stage of economic development. Various countries differ in terms of their culture and traditions as a result sustainability schemas and practices adopted in such countries deviate a lot (Witt and Redding 2013). Similarly, Aguilera et al. (2007) contended that 'because business organizations are embedded in different national systems, they will experience divergent degrees of internal and external pressures to engage in social responsibility initiatives' (p. 836). Consumers today prefer ethical and green products that reflect their own values and beliefs (Hennigs et al. 2013). According to Havas Media sustainable consumption research performed with 20,000 customers in ten countries, disclosed that 86% of buyers examined sustainability aspects while making purchase decisions. It was also revealed that 80% of consumers under the age of 35 wanted to go for sustainable option (Hashmi 2017). The concept of sustainability encompasses seeking wellness, a better quality of life, and a sense of responsibility toward the community (Pavione et al. 2016). Millennial aspire to rationalize their luxury buying through contributing to social well-being. The concept of luxury across societies has evolved from 'material' to 'immaterial' (Hashmi 2017). Figure 5 depicts the factors leading to changes in societal values that drive luxury companies to develop new business models through 'LSCV lens.'

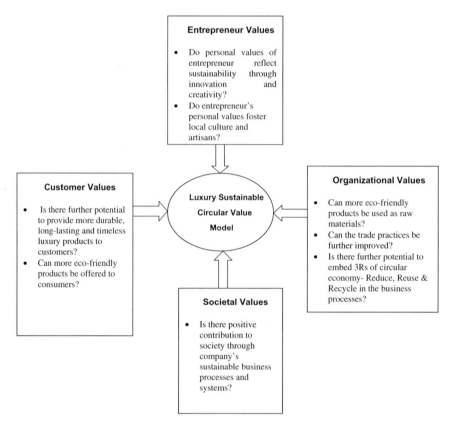

Fig. 6 LSCV model—a diagnosis tool. *Source* Designed by authors

9   LSCV—A Tool for Diagnosis

The LSCV model can be used as a diagnosis tool to examine how luxury brands contribute toward creating sustainable circular luxury value (Fig. 6).

10   Natweave Textile Studio (Case Company)

> I believe in blending time-tested traditional techniques with contemporary design trends to create sustainable luxury products.
>
> —Subhabrata Sadhu, Founder, Sadhu

Sadhu calls himself a wandering designer who is willing to climb any mountain to get the finest fleece and meet the best artisan. He has keen desire to preserve dying weaves, patterns, and skills as well as to enable and empower their custodians. He has worked with few Non-Governmental organizations to execute his ideas, assisting rural artisans design contemporary items. It has led many craftsmen to become entrepreneurs and preserve their craft. He has travelled abroad extensively which has helped him to better understand taste, preferences, and cultures of different parts of the world. It has enabled him to create modern, cosmopolitan, and luxurious look for his brand. He has been a finalist in the Young Creative Entrepreneur Award 2010, British Council of India.

Sadhu collection of scarves and wraps exhibits the 2000-year-old legacy of Pashima as narrated by Kashmiri weavers. His collection transforms traditional weaves into unique offerings by using creative and innovative interplay of colors, designs, and patterns. His range of shawls and scarves is completely unadulterated and pure with no mechanized processes used, making them extremely light, soft, and warm. Handmade production process makes his collection much more expensive than the machine-made cashmere.

Sadhu, as a textile designer, fell in love with the Pashmina fiber, popularly known as '*Fiber for royals and emperors*,' when he first encountered it. *He* was mesmerized by the extraordinary beauty, culture, heritage, craftsmanship, and architecture of Kashmir. Natweave was originated in 2009 with a passion to revive the dying Indian Pashmina industry. Indian Pashmina shawls are renowned around the world for the unique way in which they are being prepared right from sorting of raw material to the finished product. Only the artisans in the Kashmir valley only have the know-how to produce these shawls. This art has been passed to the Kashmiris from one generation to another as a legacy (Yaqoob et al. 2012). However, political unrest in the valley and infrastructural issues have increased manufacturer's preference for low-cost, low-quality, machine-made Pashmina products. This has resulted in 90% of women spinners quit this job in the last two decades. Therefore, with a yearning to protect the rare art form as well as curators of this art—indigenous Kashmiri spinners and weavers, Sadhu decided to organize the artisans of this sector and showcase centuries old timeless Kashmiri art to the world.

Sadhu creates his weaves from Pashmina fiber. It is the down fiber or undercoat obtained from special goat known as Capra Hircus/Pashmina goat/Changra goat reared in its natural habitat in Changthang plateau in the Kashmir region (Fig. 7). Changra goats are nurtured by Changpa tribe at high altitude where the temperature during winter was as low as—104 °F (Shakyawar et al. 2013). Each goat produces about 80 g of pure Pashmina, making this fiber a rare commodity. Changthang region of India produces the finest and best quality of Pashmina wool across the globe. The average fiber diameter of Changra goats is 11–13 μm[1] (whereas human hair is 200 μm) which makes it very soft. The name Pashmina is derived from Persian word pashm meaning soft gold, the king of fibers. Pashmina, globally popular as 'Cashmere' is a fine luxury fiber which commands a special position among all the fibers of animal origin due to its warmth, softness, lightness, elegance, timelessness, and its superior ability to absorb dyes and moisture (Yaqoob et al. 2012).

---

[1] http://Pashminashawls.co.in/Pashmina-ladakh (accessed September 30, 2017).

Fig. 7 Pashmina is fine, soft, and deliciously warm fabric made from the fleece of Pashmina goats. *Source* Sadhu's Website (http://www.sadhucollective.com)

11   Sadhu's Value Chain

Based on the information collected from the Kashmiri artisans through semi-structured interview, this study determines the process adopted by them to create Pashmina shawl from the raw Pashmina fiber. The process is broadly divided into four heads:

A. *Pre-spinning*: During the spring season, Pashmina is harvested, when the goat naturally shed their undercoat. It is done manually by combing which is followed by machine dehairing.[2] Then, impurities like dust, vegetable matter are removed from raw Pashmina which is followed by glueing[3] and further cleaning to remove all traces of crushed rice powder.

B. *Spinning*: Next step is spinning which is being carried out on a spinning wheel termed *chakra* (Fig. 8). It is a skillful operation carried by local Kashmiri women who are maestro in spinning the wool into thinnest yarn with their charkhas. Nowadays, with the advancement of technology, Pashmina yarn is also spun in machine by mixing nylon fiber. The nylon portion of the yarn is then dissolved by using commercial grade hydrochloric acid. This process weakens the fabric (reducing its life span) and is not considered eco-friendly. However, Sadhu still

---

[2]Dehairing means separating of pashmina fibers (undercoat) from course outer coat.

[3]Glueing means application of glueing material to Pashmina to provide extra strength and softness to the fiber. Pounded powdered rice is used as glueing material.

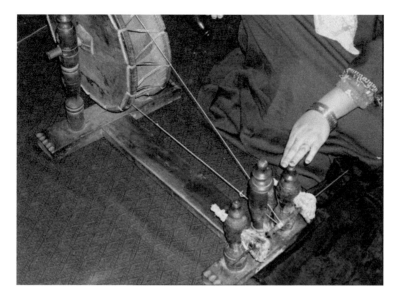

Fig. 8 Hand spinning on chakra. *Source* Sadhu

Fig. 9 Handloom. *Source* Sadhu's Website (http://www.sadhucollective.com)

uses traditional handmade spinning process for producing world-class sustainable Pashmina products.

C. *Weaving*: Pashmina yarn is weaved into shawl in a special type of handloom (Fig. 9). Weaving is performed by experienced and skilled artisans. After weaving, the fabric is *hand massaged* for reducing the tension inserted during spinning and weaving process.

Fig. 10 Finished product—Pashmina scarf. *Source* Sadhu's Website (http://www.sadhucollective. com)

D. *Finishing*: Once weaving is done, fabric is washed in cold water with powdered soap nut, *reetha*.[4] Then, the washed fabric is clipped to remove any superficial flaw on the surface. Again, the fabric is washed in running water and then dried. The fabric is then rolled and left stretched for few days and then ironed. Sadhu with the help of Kashmiri artisans develops his unique range of Pashmina shawls and scarves by using different types of weaves, hand embroidery, block printing, etc. (Fig. 10).

Natweave presently works with 30 craftsmen who work from their homes in Kashmir valley as Pashmina is primarily a cottage industry where artisans operate from their households. Sadhu follows *fair trade practices*. Each artist earns around $10 per day. They work for around 25 days in a month making at an average $250 per month. Mostly, there are two weavers working in a family (like father and son). In June, during the peak summer season, weavers estimate the demand for the Pashmina products and buy fiber from the cooperatives. Sadhu enriches and empowers the weavers to work as entrepreneurs. He explained:

> I don't want artisans to work as contractors, I want them to work as entrepreneurs in their own rights. The artisans have full liberty to work for others too.

Due to the finest quality, exclusivity of designs and eco-friendly fabric, high demand for Sadhu's sustainable range of luxury products come from Germany, France, UK, USA, Japan, Uruguay, Chile, Australia, and South Africa. Around 60% of demand for Sadhu's products comes from USA. India comprises of around 5% his market. Customers mostly above the age of 35 buy his products either for themselves or for gifting. The wholesale price range of Sadhu's shawls and scarves starts from $100 and goes up to $400 depending upon the size, weave, weight, technique, and design. Sadhu uses eco-friendly shopping bag, made of natural fiber, cotton silk for packaging. It allows the art piece to breathe and have a longer life span. His logo is

---

[4]Reetha is special soap made from herbal ingredients.

also block printed on the cloth bag. Company has been consistently growing over the years.

Sadhu believes that, *'To own a Sadhu scarf or shawl is to own a piece of the world's most premium Pashmina, woven by hand on the looms of time in designs of timeless appeal.'*

12 Natweave Textile Studio and Luxury Sustainable Circular Value (LSCV) Model: Diagnosis

The concept of luxury and sustainability is antithetical to each other. However, recently sustainability 'is rapidly becoming an issue of critical importance for designers and society as a whole' (Wahl and Baxter 2008). Various luxury companies are now emphasizing on sustainability. For example, Prada is using variety of biodegradable natural fibers in place of polyester. Stella Mc Cartney uses no leather in her luxury collection. Her 70% products are *handmade,* recyclable, and reusable.

Sustainability is not only about sustainable environment and using eco-friendly materials but also about creating an integrative sustainable value chain and promoting and empowering people to improve their quality of life. Natweave is one such company which is committed to conserve the 2000 years old legacy of Pashmina and sustainable methods of production. It integrates the values of four stakeholders—entrepreneur, organization, customers, and society. Figure 11 shows a brief summary of the strategies and practices presently followed by Natweave for creation of luxury sustainable circular value (LSCV). Right from procurement of raw material to spinning and weaving to finishing, Sadhu's luxury offerings are completely handmade to ensure that the unique fabric maintains its natural softness and warmth and thus 100% sustainable. Sadhu explained:

> The brand is perfect tribute to the dying art of hand weaving and hand spinning, empowering Kashmiri weavers and artisans, while preserving and promoting the traditional production techniques of the region believed to be two thousand years old.

Sadhu has obtained *Geographical Indication* (GI) registration from Government of India. GI mark has been imprinted in the form of label after ensuring originality of fabric, fineness of thread, spinning method, and weaving technology.

13 Natweave Textile Studio: Road Ahead

Sadhu could make further efforts to integrate the values of all four stakeholders—entrepreneur, organization, customers, and society simultaneously. Based on questions in Fig. 6,[5] an attempt is made to provide detailed analysis of the practices Natweave

---

[5]This analysis is limited to the questions mentioned in Fig. 6. It is also susceptible to some biasness from the authors while evaluating each stakeholder.

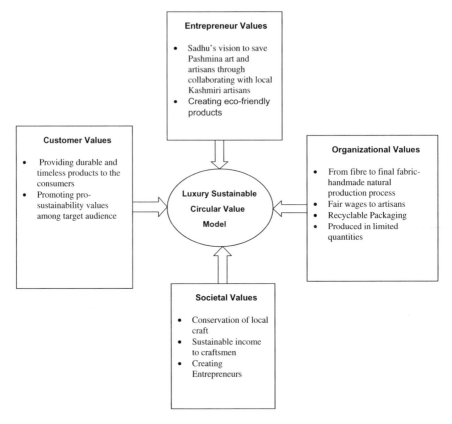

Fig. 11 Strategies and practices followed by Natweave for creation of luxury sustainable circular value (LSCV) model. *Source* Designed by authors

could follow (Fig. 12) to maximize luxury sustainable circular value (LSCV) creation.

13.1 Entrepreneur

If a leader possesses pro-environmental values and recognizes his/her firm as committed toward environment, it is more likely that firm will embrace environment-friendly strategies and innovations (Branzei et al. 2000). As the key reason behind establishment of Natweave is to create value for society and environment, it could further look for newer methods to blend technology and innovation and create world-class, sustainable, Pashmina product range through collaborating with more Kashmiri artisans.

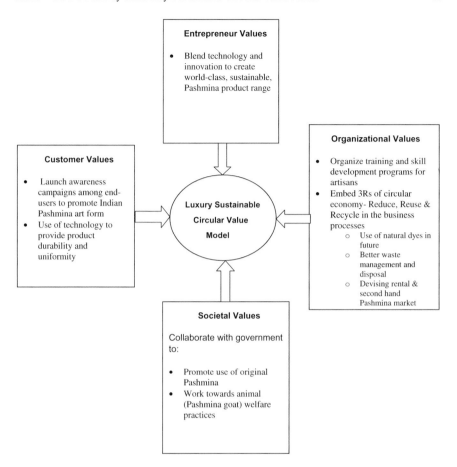

Fig. 12 Strategies and practices Natweave could follow to maximize luxury sustainable circular value (LSCV) creation. *Source* Designed by authors

13.2 Organization

An organization can find solution to social and environmental problems either through internal development or acquisition of new capabilities (Hart 2005, 2007). Natweave can further initiate effective use of resources through organizing training and development programs for artisans. It can encourage artisans to use new methods and processes to produce innovative weaves. Common natural colors of Pashmina are gray, brown black, and white. For providing vibrant shades to the fiber, skin-friendly AZO-free chemical dyes are used which has lesser impact on environment. In future, to further minimize the impact on environment in terms of soil and water pollution, Natweave may either use natural dyes or start in-house dying facility at weaver's place, where water consumption and chemical usage can be controlled with close supervision leading to minimization of waste as well as pollution prevention. It may also look for alternative solutions like storing the wastewater (with chemicals) in a tank and further treating it with some solution which could neutralize its harmfulness. In addition, renting and second-hand market concepts can be encouraged to extend the typical life cycle of Pashmina shawls through several users.

13.3 Customers

For sustainable consumption, consumers rely on businesses to provide products that have been developed in a sustainable manner. They look for assistance on which goods to select, how to use them properly, and how to ensure that they are reused or recycled. Netweave can make efforts to educate consumers about the process through which Pashmina shawl is created. Even today, not many people are aware about the environmental and social aspects of the *Cashmere* industry. They could be motivated to purchase original Pashmina rather than machine-made replicas to save the art form as well as the artists. Also, through the use of technology and innovation, wider range of sustainable options could be created for the consumers.

13.4 Society

As Pashmina handloom industry is a major source of income for the state of Jammu and Kashmir, government has made several efforts to revive this industry. Netweave can collaborate with government to promote the use of original Pashmina rather than cheap machine-made replicas to ensure survival of this rare art and artisans. They can together work toward better management of Pashmina goats, their shed, food, etc.

14 Conclusion

Value creation and how a designer can play a vital role in creating new value is one of the challenges in luxury industry. Players operating in luxury industry must be able to design and envisage value models that link consumers' requirements, improve supply chain and operational practices, follow innovative practices to add value to sustainable development. As a responsible marketer, luxury brands should not only consider economic values, but also on societal, cultural, ethical, and environmental values. In this study, authors developed a luxury sustainable circular value (LSCV) model through assimilating the values of four stakeholders—entrepreneur, organization, customers, and society. Firstly, any luxury entrepreneur on the basis of his personal values, knowledge, motivation, and stimulus can create sustainable circu-

CIRCULAR	SERVITIZATION	SUFFICIENCY
Create value from waste • Through better waste management and disposal	**Functionality over Ownership** • Experiential value over possessions	**Encourage effective resource use** • Through training and development of artisans
Closed Loop Models • Help create newer jobs for society	**Repair & Warranty** • New life through luxury product maintenance	**Demand Management** • Mitigation of surplus stock
Reuse • Devising second hand Pashmina market • Through Inheritance	**Renting & Leasing** • Extended lifecycle through prolonged use phase • Make luxury accessible	**Co-creation** • Offer customers opportunity to choose their own colors, weave, size, design, etc.

Fig. 13 Natweave's business model framework to create LSCV. *Source* Designed by Authors (Adapted from Circle Economy and Sitra 2015)

lar value. Entrepreneur should have vision to create sustainable development. He should visualize to resuscitate the local art and culture and must have craving for economic, environmental, and social profits. Secondly, consumers of luxury brands can contribute toward creating value in terms of their functional, financial, personal, and interpersonal values. This new form of sustainable circular consumption should lie in the center of consumer value system and consumers' emphasis should be more on impact on community rather than self-enhancement values. Thirdly, organization vision of sustainability and circularity provides direction to the members of organization regarding existing practices, organizational priorities, technological innovations, and business models. Organizations have to rethink the way they do business through the vision of a reformative, circular economy that creates value for the organization. Last but not the least, society's value toward sustainable orientation can influence how we educate and provide values to our next generation. Societies need to change their beliefs and value system toward sustainability and look for better quality of life that can be achieved through sustainable circularity concept. Further, the paper examined how 'Natweave Textile Studio' creates value by focusing on above-mentioned four stakeholders. Old luxury principle where long-lasting, superior quality, hand-crafted products are rarely purchased and kept for longer period of time before disposal (Hennigs et al. 2013; Okonkwo 2007) holds true for Pashmina. In future, Sadhu can use business model framework (Fig. 13), provided by Circle Economy and Sitra (2015) to further create LSCV for its organization.

References

Achabou, M. A., & Dekhili, S. (2013). Luxury and sustainable development: Is there a match? *Journal of Business Research, 66*(10), 1896–1903.

Agle, B. R., Mitchell, R. K., & Sonnenfeld, J. (1999). Who matters to CEOs? An investigation of stakeholder attributes and salience, corporate performance, and CEO values. *Academy of Management Journal, 42,* 507–525.

Aguilera, R. V., Rupp, D. E., Williams, C. A., & Ganapathi, J. (2007). Putting the S back in corporate social responsibility: A multilevel theory of social change in organizations. *Academy of Management Review, 32*(3), 836–863.

Bansal, P. (2003). From issues to actions: The importance of individual concerns and organizational values in responding to natural environmental issues. *Organization Science, 14*(5), 510–527.

Bansal, P., & Roth, K. (2000). Why companies go green: A model of ecological responsiveness. *Academy of Management Journal, 43*(4), 717–736.

Baron, R. A., & Shane, S. (2007). *Entrepreneurship: A process perspective* (2nd ed.). Cincinnati, OH: Thomson-Southwestern.

Beard, N. D. (2008). The branding of ethical fashion and the consumer: A luxury niche or mass-market reality? *Fashion Theory, 12*(4), 447–468.

Bearden, W. O., & Etzel, M. J. (1982). Reference group influence on product and brand purchase decisions. *Journal of Consumer Research, 9*(2), 183–194.

Belk, R. W. (1988). Possessions and the extended self. *Journal of Consumer Research, 15,* 139–168.

Bemporad, R., Hebard, A., Bressler, D. (2012). Re: Thinking consumption, consumers and the future of sustainability. BBMG, GlobeScan and SustainAbility. Retrieved on 21 Sept, 2017 form http://www.globescan.com/news-and-analysis/press-releases/press-releases-2013/98-press-releases-2013/257-aspirational-consumers-unite-style-sustainability-to-shape-market-trends.html.

Bendell, J., Kleanthous, A. (2007). Deeper luxury. Retrieved on 27 Sept, 2017 from http://www.wwf.org.uk/deeperluxury/_downloads/DeeperluxuryReport.pdf.

Berson, Y., Oreg, S., & Dvir, T. (2008). CEO values, organizational culture and firm outcomes. *Journal of Organizational Behavior, 29,* 615–633.

Berthon, P., Leyland, F. P., Parent, M., & Berthon, J. P. (2009). Aesthetics and ephemerality: Observing and preserving the luxury brand. *California Management Review, 52*(1), 45–66.

Branzei, O., Vertinsky, I., Zietsma, C. (2000). From green-blindness to the pursuit of eco-sustainability: An empirical investigation of leader cognitions and corporate environmental strategy choices. In *Academy of Management Proceedings & Membership Directory,* C1–C6.

Brun, A., Caniato, F., Caridi, M., Castelli, C., Miragliotta, G., Ronchi, S., et al. (2008). Logistics and supply chain management in luxury fashion retail: Empirical investigation of Italian firms. *International Journal of Production Economics, 114,* 554–570.

Burns, B. (2010). Re-evaluating obsolescence and planning for it. In T. Cooper (Ed.), *Longer lasting products: Alternatives to the throwaway society* (pp. 39–60). Gower: Farnham.

Carrier, J. G., & Luetchford, P. (2012). *Ethical consumption: Social value and economic practice.* New York: Berghahn Books.

Cervellon, M. C., & Shammas, L. (2013). The value of sustainable luxury in mature markets: A customer-based approach. *Journal of Corporate Citizenship, 52,* 90–101.

Chapman, M. (2000). When the entrepreneur sneezes, the organization catches a cold: A Practitioner's perspective on the state of the art in research on the entrepreneurial personality and the entrepreneurial process. *European Journal of Work and Organizational Psychology, 9*(1), 97–101.

Chapple, W., & Moon, J. (2005). Corporate social responsibility (CSR) in Asia: A seven-country study of CSR web site reporting. *Business and Society, 44*(4), 415–441.

Charles, G. (2010). Ethics come into fashion. *Marketing.* Feb 24, 16.

Circle Economy and Sitra (2015). Service-based business models & circular strategies for textiles. Retrieved on Nov 12, 2017 from http://www.slideshare.net/SitraEkologia/servicebased-business-models-circular-strategies-for-textiles.

Cooper, T. (2005). Slower consumption: Reflections on products' life spans and the 'throwaway society. *Journal of Industrial Ecology, 9*(1–2), 51–67.

Corman, J., Perles, B., & Vancini, P. (1988). Motivational factors influencing high-technology entrepreneurship. *Journal of Small Business Management, 26*(1), 36–42.

Crals, E., Vereeck, L. (2004). *Sustainable entrepreneurship in SMEs: Theory and practice.* Paper presented at the 3rd Global Conference in Environmental Justice and Global Citizenship, Copenhagen, Denmark.

Cvijanovich, M. (2011). Sustainable luxury: Oxymoron? Lecture in Luxury and Sustainability. Lausanne, July 2011. Retrieved on Aug 25, 2017 form http://www.mcmdesignstudio.ch/files/Guest%20professor%20Lucern%20School%20of%20Art%20%20and%20Design.pdf.

Davies, I. A., Lee, Z., & Ahonkai, I. (2012). Do consumers care about ethical luxury? *Journal of Business Ethics, 106*(1), 37–51.

De Barnier, V., Rodina, I., Valette-Florence, P. (2006). Which luxury perceptions affect most consumer purchase behavior? A cross cultural exploratory study in France, the United Kingdom and Russia. In *International Marketing Trends Conference*. Paris.

Dees, G. J. (2001). *The meaning of 'social entrepreneurship'*. Stanford, CA: Stanford Graduate Business School.

Djelic, M. L., & Ainamo, A. (1999). The coevolution of new organizational forms in the fashion industry: A historical and comparative study of France, Italy and the United States. *Organizational Science, 10*, 622–637.

Doval, J., Singh, E. P., & Batra, G. S. (2013). Green buzz in luxury brands. *Review of management, 5*, 1–23.

Dubois, B., Czellar, S., Laurent, G. (2005). Consumer segments based on attitudes toward luxury: Empirical evidence from twenty countries. *Marketing Letters, 16* (2), 115–128.

Ducrot-Lochard, C., & Murat, A. (2011). *Luxe et développement durable: La nouvelle alliance.* Paris: Eyrolles.

Elkington, J. (1999). *Cannibals with forks: the triple bottom line of 21st century business*. Oxford: Capstone.

Ellen MacArthur Foundation (2013). Towards the circular economy Vol. 1: An economic and business rationale for an accelerated transition. Retrieved on 25 Oct, 2017 on https://www.ellenmacarthurfoundation.org/assets/downloads/publications/Ellen-MacArthur-Foundation-Towards-the-Circular-Economy-vol.1.pdf.

European Multistakeholder Forum on CSR (2004). European Multistakeholder Forum on CSR: Final results & recommendations. Retrieved on 31 Oct, 2017 from http://www.indianet.nl/EU-MSF_CSR.pdf.

Fagenson, E. A. (1993). Personal value systems of men and women entrepreneurs versus managers. *Journal of Business Venturing, 8,* 409–431.

Feng, W. J., Mao, Y. R., Chen, H., & Chen, C. (2007). Study on development pattern of circular economy in chemical industry parks in China. *Xiandai Huagong/Modern Chemical Industry, 27*(3), 7–10.

Fletcher, K. (2008). *Sustainable fashion & textiles: Design journeys*. London: Earthscan.

Gardetti, M. A., Girón, M. E. (2017). Sustainable luxury and social entrepreneurship: Stories from the pioneers. UK: Greenleaf Publishing.

Gardetti, M. A., Torres, A. L. (2013). Sustainability in fashion and textiles: Values, design, production and consumption. *Management of Environmental Quality: An International Journal, 24* (4).

Genovese, A., Acquaye, A. A., Figueroa, A., & Koh, S. L. (2017). Sustainable supply chain management and the transition towards a circular economy: Evidence and some applications. *Omega, 66*, 344–357.

Grail Research (2010). 'Green—The new color of luxury: Moving to a sustainable future. Retrieved on Oct 31, 2017 from http://www.grailresearch.com/pdf/Blog/2010-Dec-Grail-Research-Green-The-New-Color-of-Luxury_244.pdf.

Hart, S. L. (2005). *Capitalism at the crossroads: The unlimited business opportunities in solving the world's most difficult problems.* Upper Saddle River: Wharton School Publishing.

Hart, S. L. (2007). *Capitalism at the crossroads: Aligning business, earth, and humanity* (2nd ed.). Upper Saddle River, N.J.: Wharton School Publishing.

Hart, S. L., Milstein, M. B. (2003). Creating sustainable value. The Academy of Management Executive.

Hashmi, G. (2017). Redefining the essence of sustainable luxury management: The slow value creation model. Sustainable management in luxury (pp. 3–27). Singapore: Springer.

Heck, P. (2006). Circular economy related international practices and policy trends: Current situation and practices on sustainable production and consumption and international circular economy development policy summary and analysis. World Bank Report.

Hemingway, C. A. (2005). Personal values as a catalyst for corporate social entrepreneurship. *Journal of Business Ethics, 60*(3), 233–249.

Hennigs, N., Wiedmann, K. P., Klarmann, C., & Behrens, S. (2013). Sustainability as part of the luxury essence: Delivering value through social and environmental excellence. *Journal of Corporate Citizenship, 52,* 25–35.

Henninger, C. E., Alevizou, P. J., & Oates, C. J. (2016). What is sustainable fashion? *Journal of Fashion Marketing and Management, 20*(4), 400–416.

Henninger, C. E., Alevizou, P. J., Tan, J. L., Huang, Q., & Ryding, D. (2017). Consumption strategies and motivations of Chinese consumers: The case of UK sustainable luxury fashion. *Journal of Fashion Marketing and Management: An International Journal, 21*(3), 419–434.

Hood, J. N. (2003). The relationship of leadership style and CEO values to ethical practices in organizations. *Journal of Business Ethics, 43,* 263–273.

Hope, K. (2015). The clothing firms designing clothes that last forever. BBC News. Retrieved on 12 Oct, 2017 from http://www.bbc.com/news/business-34984836.

Isaak, R. (1998). *Green logic: Ecopreneurship, theory and ethics.* Sheffield, UK: Greenleaf Publishing.

Jain, S., Khan, M. N., & Mishra, S. (2015). Factors affecting luxury purchase intention: A conceptual framework based on an extension of the theory of planned behavior. *South Asian Journal of Management, 22*(4), 136–163.

Joy, A., Sherry, J. F., Venkatesh, J. A., Wang, J., & Chan, R. (2012). Fast fashion, sustainability, and the ethical appeal of luxury brands. *Fashion Theory, 16*(3), 273–296.

Kale, G.O. & Öztürk, G. (2016). The importance of sustainability in luxury brand management. *Intermedia International e-Journal, 3*(1).

Kapferer, J. N. (2010) All that glitters is not green: The challenge of sustainable luxury. Retrieved on Aug 25, 2017 from www.europeanbusinessreview.com/?p=2869.

Kapferer, J. N. (2013). All that glitters is not green: The challenge of sustainable luxury. The European Business Review. Retrieved on Aug 11, 2017 from http://politicalanthropologist.com/?p=2869.

Kapferer, J. N., & Bastien, V. (2012). *The luxury strategy: Break the rules of marketing to build luxury brands.* London: Kogan Page.

Karp, D. G. (1996). Values and their effect on pro-environmental behavior. *Environment and Behavior, 28*(1), 111–133.

Kendall, J. (2010). Responsible luxury: A report on the new opportunities for business to make a difference. Retrieved on Oct 25, 2017 from www.cibjo.org/download/responsible_luxury.pdf.

Kim, A. J., & Ko, E. (2012). Do social media marketing activities enhance customer equity? An empirical study of luxury fashion brand. *Journal of Business Research, 65*(10), 1480–1486.

Kim, K. H., Ko, E., Xu, B., & Han, Y. (2012). Increasing customer equity of luxury fashion brands through nurturing consumer attitude. *Journal of Business Research, 65*(10), 1495–1499.

Kleanthous, A. (2011). Simply the best is no longer simple. The Raconteur—Sustainable Luxury, July 2011.

Krishnan, V. R. (2001). Value systems of transformational leaders. *Leadership & Organization Development Journal, 22*(3), 126–132.

Meglino, B. M., & Ravlin, E. C. (1998). Individual values in organizations: Concepts, controversies, and research. *Journal of Management, 24*(3), 351–389.

Möller, K. (2006). Role of competence in creating customer value: A value-creation logic approach. *Industrial Marketing Management, 35,* 913–924.

Murray, A., Skene, K., & Haynes, K. (2017). The circular economy: An interdisciplinary exploration of the concept and application in a global context. *Journal of Business Ethics, 140,* 369–380.

Nair, C. (2008). The last word. Luxury goods: Ethics out of fashion. Ethical Corporation. Retrieved on Nov 15, 2017 from http://www.global-inst.com/ideas-for-tomorrow/article/2008/the-last-word-luxury-goods-ethics-out-of-fashion.

Nonis, S., & Swift, C. O. (2001). Personal value profiles and ethical business decisions. *Journal of Education for Business, 76*(5), 251–256.

Okonkwo, U. (2007). *Luxury fashion branding: Trends, tactics, techniques*. Basingstoke: Palgrave Macmillan.

Olorenshaw, R. (2011). Luxury and the recent economic crisis. *Vie & Sciences Economiques, 188,* 72–90.

Pavione, E., Pezzetti, R., & Dallâ, M. (2016). Emerging competitive strategies in the global luxury industry in the perspective of sustainable development: The case of Kering Group. *Management Dynamics in the Knowledge Economy Journal, 4*(2), 241–261.

Pitt, J. (2011). Beyond sustainability? Designing for a circular economy. Retrieved on Oct 24, 2017 from http://www.ort.org/uploads/media/10th_Hatter_booklet.pdf.

Ramchandani, M., & Coste-Maniere, I. (2012). Asymmetry in multi-cultural luxury communication: A comparative analysis on luxury brand communication in India and China. *Journal Global Fashion Marketing, 3*(2), 89–97.

Rohan, M. J. (2000). A rose by any name? The values construct. *Personality and Social Psychology Review, 4*(3), 255–277.

Rokeach, M. (1973). *The nature of human values*. New York: The Free Press.

Schaper, M. (2002). The essence of ecopreneurship. *Greener Management International, 38,* 26–30.

Schick, H., Marxen, S., & Freimann, J. (2002). Sustainability issues for start-up entrepreneurs. *Greener Management International, 38,* 59.

Seringhaus, F. H. R. (2002). Cross-cultural exploration of global brands and the internet. In *18th Annual IMP Conference*. Dijon.

Shakyawar, D. B., Raja, A. S. M., Kumar, A., Pareek, P. K., & Wani, S. A. (2013). Pashmina fibre—Production, characteristics and utilization. *Indian Journal of Fibre & Textile Research, 38,* 207–214.

Shane, S., & Venkataraman, S. (2000). The promise of enterpreneurship as a field of research. *The Academy of Management Review, 25*(1), 217–226.

Shen, M. H. (2007). *Recourse and environment economics*. Beijing, China: China Environmental Science Press.

Siegle, L. (2009). Cardiff's online shoppers are worst offenders for binning unworn clothes. The Guardian. Retrieved on Oct 7, 2017 from https://www.theguardian.com/environment/ethicallivingblog/2009/jan/28/clothes-landfill-global-cool-co2-emissions.

Spangenberg, J. (2001). Sustainable development. From catchwords to benchmarks and operational concepts. In M. Charter & U. Tischner (Eds.), *Sustainable solutions, developing products and services for the future* (pp. 24–47). Sheffield: Greenleaf.

Stead, J. G., & Stead, W. E. (2009). *Management for a small planet* (3rd ed.). Armonk: M E Sharp.

Steinhart, Y., Ayalon, O., & Puterman, H. (2013). The effect of an environmental claim on consumers? Perceptions about luxury and utilitarian products. *Journal of Cleaner Production, 53,* 277–286.

Torelli, C. J., Monga, A. B., & Kaikati, A. M. (2012). Doing poorly by doing good: Corporate social responsibility and brand concepts. *Journal of Consumer Research, 38*(5), 948–963.

Valente, M. (2012). Theorizing firm adoption of sustaincentrism. *Organization Studies, 33*(4), 563–591.

Vigneron, F., & Johnson, L. W. (2004). Measuring perceptions of brand luxury. *Journal of Brand Management, 11*(6), 484–506.

Wahl, D. C., & Baxter, S. (2008). The designer's role in facilitating sustainable solutions. *Design Issues, 24*(2), 72–83.

Waldman, D. A., Siegel, D. S., Javidan, M. (2004). CEO transformational leadership and corporate social responsibility. Retrieved on Oct 7, 2017 from http://www.economics.rpi.edu/workingpapers/rpi0415.pdf.

Wang, J., & Wallendorf, M. (2006). Materialism, status signaling and product satisfaction. *Journal of Academy of Marketing Science, 34*(4), 494–505.

Wiedmann, K. P., Hennigs, N., & Siebels, A. (2007). Measuring consumers' luxury value perception. *Academy of Marketing Science Review, 11*(7), 1–21.

Wiedmann, K. P., Hennigs, N., & Siebels, A. (2009). Value-based segmentation of luxury consumption behavior. *Psychology and Marketing, 26*(7), 625–651.

Witt, M. A., & Redding, G. (2013). Asian business systems: institutional comparison, clusters and implications for varieties of capitalism and business systems theory. *Socio-Economic Review, 11*(2), 265–300.

World Commission on Environment and Development (WCED). (1987). *Our common future.* Oxford, UK: Oxford University Press.

Wu, J. S. (2005). *New circular economy.* Beijing: Tsinghua University Press.

Yaqoob, I., Sofi, A. H., Wani, S. A., Sheikh, F. D., & Bumla, N. A. (2012). Pashmina shawl—A traditional way of making in Kashmir. *Indian Journal of Traditional Knowledge, 11*(2), 329–333.

Zhu, D., & Qui, S. (2007). Analytical tool for urban circular economy planning and its preliminary application: a case of Shangai. *Urban Ecological Planning, 31*(3), 64–70.

Sheetal Jain has more than 13 years of experience in academia and industry. She is the founder & CEO of Luxe Analytics, a specialized luxury market intelligence and strategic advisory firm. She holds PhD in luxury marketing from Aligarh Muslim University. Her research papers have been published in refereed national journals and international journals like Journal of Asia Business Studies; South Asian Journal of Management, Journal of Fashion Marketing and Management, etc. She is in the review panel of various refereed international journals. She has made paper presentations in international/ national conferences including IIM Ahmadabad; IIM Kashipur; IIT Delhi, etc. She is also a visiting faculty to leading academic institutions such as IMT Ghaziabad, IILM, Luxury Connect Business School, and Pearl Academy.

Sita Mishra has more than 20 years of experience in industry as well as academics. At present, she is working as an Associate Professor in IMT-Ghaziabad, India. She has qualified National Eligibility Test for Faculty (NET), conducted by the University Grants Commission (UGC), Government of India, New Delhi, in 1993 and is also a Junior Research Scholarship holder, granted by UGC. She has published more than 40 research papers in international/national journals/book chapters and presented papers in international/national conferences, besides being on the reviewing board of a few journals.

Cradle to Cradle®—Parquet for Generations: Respect Natural Resources and Offers Preservation for the Future

Ansgar Igelbrink, Albin Kälin, Marko Krajner and Roman Kunič

Abstract Indoor air quality matters in regard of health risks of indoor exposure to particulates. Small particulates indoor are 3–8 times higher than outdoor (Heimlich 2008). The air quality indoor suffers. Industrial products such as building materials, paints, furniture, textiles, flooring, and electronics are off-gazing and in general incorporate toxic ingredients. The environmental and human toxicology quality is a key factor for a healthy living environment. Health risks of indoor exposure to particulates matter in regard to quality of industrial produced products. Wood will certainly stay as one of the leading and the most preferable construction material in the future due to its environmental, local availability, and aesthetic characteristics. Nowadays, using wood in architecture is very fashionable. Research and developments in wood production as well as in wood construction will strongly form the future of sustainable development practically in all parts of the planet Earth. Wood today is trendy, fashionable, and one of the most accessible materials and has an aesthetic view with a natural look and a visual attractiveness, together with the smell, sound, and touch; natural wood is perceived as luxurious. Development could be seen also in façades, inside and other surfaces of modern structures which are increasingly being used. Sustainable luxury products incorporate extraordinary aesthetics, handle, care, function and in addition, to be sustainable need to be safe for humans, society, and the environment. Resources and natural resources are scare and need to be protected in changing the design of the products we use according Cradle to Cradle® principle 'Remaking the way we make things' and 'Towards a circular economy.' For companies, this implies entrepreneurship to tackle the large impact in change of behavior, culture, marketing and business models in closing the loop, and taking the goods back from the user. The case study, Cradle to Cradle® (McDonough and Braungart

A. Igelbrink
Bauwerk Parkett AG, Neudorfstrasse 49, 9430 St. Margrethen, Switzerland

A. Kälin (✉) · M. Krajner · R. Kunič
EPEA Switzerland GmbH, Seestrasse 119, 8806 Baech, Switzerland
e-mail: kaelin@EPEAswitzerland.com

M. Krajner
Faculty of Civil and Geodetic Engineering, University of Ljubljana, Jamova 2, 1000 Ljubljana, Slovenia

© Springer Nature Singapore Pte Ltd. 2019
M. A. Gardetti and S. S. Muthu (eds.), *Sustainable Luxury*,
Environmental Footprints and Eco-design of Products and Processes,
https://doi.org/10.1007/978-981-13-0623-5_5

83

2002)—Parquet for Generations (Bauwerk Parkett 2017)—Respect Resources and Preservation for the Future, illustrates a successful lighthouse example from industry.

Keywords Parquet · Cradle to Cradle · Sustainable development · Living and working environment · Circular economy · Carbon dioxide

1 Introduction

We do not inherit the Earth from our Ancestors; we borrow it from our Children. Unfortunately, world's population have caused huge influence on the environment, exceeding carrying capacity of the planet Earth, and if our nowadays society and existing economy are not transformed drastically, we risk descent into polluted environment and unhealthy urban conditions, depletion of virgin materials as well as loss of precious biodiversity. Our current model of economic and social growth is stimulating this unhealthy system, and, as a consequence, we have already passed the upper limits of our Earth's capability to support us.

It is important that construction wood remains during installation and use period as pure as possible, without any environment unfriendly additives, primers, coatings, impregnations, or mixtures of different bio-products in various composite materials, which are difficult or impossible to separate after use.

Forest-based bio-products, where wood represents its main position, as a source for structure on non-structure material in the building sector are the same as passive sun gain in the environmental friendly consumption economy, the most acceptable and available non-artificial and non-toxic renewable source. Bio-based raw substances are grown by the help of sun insolation in visible and IR radiation spectrum that reaches at our planet, and its surface, in non-constant, variations are drastic for the various positions on the Earth, time of the day/year, weather, and other situations. Helps the photosynthesis development is biological growing by using CO_2 from the air, what resulted that this same quantity of CO_2. This global warming potential gas (in this case CO_2) is kept in bio-based raw material until the discharging with oxidation progression occurs (fire, rotting, bacterial and fungal decays, ingestion by animals). Wood as a bio-product is ideal for storing CO_2, especially in built houses and their bearing or non-bearing constructions, where CO_2 could be stored for longer periods counted in decades if not even centuries. Even more, after ideal closed-loop life cycle chain, bio-based elements could be easily recycled and again settled into life cycle (Kunič 2017).

Health inside environments like rooms, cabinets, offices, hotel rooms, classrooms, and other private and also public spaces is a crucial characteristic of designing proper indoor environments for living and working. Especially because people spend large part of their time inside (Kim et al. 2001; Košir et al. 2010; Gilbert et al. 2008; Kunič 2017). Furthermore, using wood products, especially natural wood, for interior surfaces and interior elements has been manifested to have very positive influence on inhabitants (Jelle 2011; Kutnar and Hill 2014; Pajek and Košir 2017;

Arkar et al. 2018; Gustavsson et al. 2006; Dovjak et al. 2012; Lakrafli et al. 2017; Paganin et al. 2017; Kitek Kuzman and Kutnar 2014).

Working with wood as a natural raw material means accepting an obligation. Natural and bio-based wood is the very precious re-growing, renewable natural substance in practically all regions. Wood is a real and modern building and architectural product, extremely useful for future. Nevertheless, by all added chemicals (glues, adhesives, fillers, protecting and preserving additives, lacquers, etc.), the resource wood becomes after the life cycle a hazardous waste. Bauwerk Parkett, a Swiss-based Company, is opening a new chapter and has adopted and developed a new technology under the name 'Silente,' which follows the sustainable development strictly according to the principle Cradle to Cradle® design. It is valued only for innovations in closed-loop substances' life cycles and guaranteed assurance in such productions that are using chemicals which are safe for humans and the environment. Unlike conventional parquets, Bauwerk Company has planned an innovative way which is suitable for generations. Thus, product will never be, even at the end of life cycle, a waste product, beside that also not to consume environmentally harmful or unnecessary energy. All raw materials, mostly natural ones, and the water resources are always used with intense care. Bauwerk Company strictly works in a fair-minded and ethically correct social responsible manner, in relation to company's workers as well as to clients and other stakeholders. On this innovative path, EPEA Switzerland has independently assisted Bauwerk with the development and execution of the sustainable vision according to Cradle to Cradle® principles. From up to 36 suppliers, all raw materials and ingredients are scientifically assessed on environmental impact, energy use and CO_2 evaluation, water treatment, and reutilization. As a result, Bauwerk Parkett products obtained under the product group name 'Silente' Gold level according to principles Cradle to Cradle certified™ certification program. Even more, all wooden floors that include the new innovative 'Silente' technology are acceptable to be returned to the Bauwerk Company after they are easily dismantled from the user's floor. With new innovation, called 'Silent-Mat', the Bauwerk's wooden floor layer could be separated without being destroyed and could be used up to three times, what means that all ingredients can be either re-conditioned or recycled for new production of final goods. The raw material lifetime will be extended from 25 years of use to 75 years of use. Those unique products consist to man, species, and the environment absolutely non-toxic substances. Thanks to this closed cycle, rather than a linear process as happens by other producers, Bauwerk conserves the valuable resource wood and acts in the interest of future generations, such as:

- defined processes of health and non-toxic raw substances and chemicals,
- consumption of all subsequent raw materials in closed life cycle,
- the evaluation of use of proper energy and CO_2 supervision,
- environmentally friendly water treatment,
- society responsibility and social fairness.

Wood is the most essential re-growing and renewable non-artificial raw material in practically all regions and is assumed to be the building material with an exceptionally bright future. Unfortunately, with hazardous chemical substances, adhesives, lacquers, the wood raw material becomes, if not before, surely after the finished life cycle, a hazardous waste.

Sustainable development under principle Cradle to Cradle® design deals with the new perspectives of a production site culture with all procedures of fabrication, use, and after use periods which are planned by transference of ethics of Nature. Thus from Cradle to Cradle®, where Nature knows material flows, unlike as by human behavior, Nature produce no waste at all (every part is totally recycled, without any waste). Nature is very effectively and cleverly implementing the right raw substances at the accepted time and on the suitable place. In the case of implementing chemicals against rotting, fire and other general resistance, innovative design should be designed with such chemicals, which are naturally acceptable and safe for bio-natural whole life periods.

2 Sustainable Development According to Cradle to Cradle® Design

The innovative idea of sustainable development according to Cradle to Cradle® design develops and defines recyclables and quality of goods. Together with all respect to difference with the existing down-cycling system, the environmental condition of the used substances stays during numerous life cycles of certain product, where only pure and safe chemicals could be used.

The goods are designed with respect to the environmental friendly system to sustain the condition of strictly all natural substances during numerous use periods, taking into account all the manufacturing procedures as well as the reutilization process. To be precise: no waste at all under all ingredients are considered only as nutrients. The right materials are integrated in defined cycles (metabolism) at the right time and place (Braungart 1992; Hawken and Lovins 1999; Gilding 2011; McDonough 1993).

The three Cradle to Cradle® design principles of sustainable development are (McDonough and Braungart 2002):

- Waste equals food,
- The energy use is only allowed in the form of renewable energy (only renewable resources),
- Diversity is allowed and is even encouraged.

Nature as a unartificial model uses sustainable developments as a principle in Cradle to Cradle® product. Sustainable developments according to innovative process in Cradle to Cradle® products reflect a new security level and differentiate model (Fig. 1).

Consumer products such as natural textiles, cosmetics, cleaning substances, detergents are planned so with care that they can be infinitely passing the biological cycle in principle to forever. The recycle to organic nutrients and promotion of biological

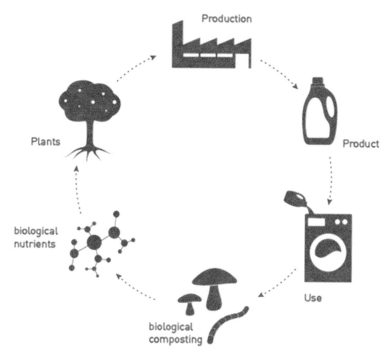

Fig. 1 Biological life cycle: organic nutrients in biological metabolism. *Source* Cradle to Cradle®

nutrients in biological life cycle, identically as in nature such as plants, are growing. The renewable raw substances are used for new products (Fig. 2).

Service Products such as TV sets, home appliances, cars, synthetic fibers, textile, etc., in other words so-called technical nutrients, are after the life cycles separated with a goal to allow the manufacture of new products or goods after finishing their original purpose. The materials after the life cycle period belonged to the manufacturer, which preserves them through well-organized system into the beginning of new technical life cycle (technical metabolism).

2.1 Differentiation Represents Quality Equal Quantity

Sustainable development under system Cradle to Cradle® design transfers to industrial systems the code 'Quality equal Quantity.' All flows of substances are planned with the goal to be valuable and suitable for the reuse and saving processes of biological and as well as technical metabolism. This methodology liberates in all views and slows down or totally reduces all harmful influences on the environment.

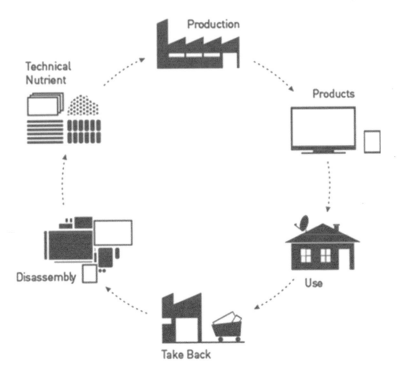

Fig. 2 Technical life cycle: technical nutrients in technical metabolism. *Source* Cradle to Cradle®

2.2 Total Beauty Design

Professor Michael Braungart defined sustainable product performance and sustainable development according to Cradle to Cradle® design as "total beauty design" (Cradle to Cradle 2017). Beauty in this case is not only based on aesthetics, but it goes beyond sustainability, beyond industrial design and beyond form follows function. "Total beauty design" incorporates a new dimension of product quality which is perceived as luxury.

3 Redefining Product Quality

International standards which are covering environmental influence of construction activities, buildings, and period of use are numerous (EN 15804 2012; ISO 14025 2009; ISO 14040 2006; ISO 14067 2013, to mention only few of them). Quality of the products, with customer service, proper customer communication, consistency in respecting its core values, and positive interactions with clients, has allowed many companies to build and hold a unique market position for the certain company and for

Fig. 3 Cradle to Cradle
logo. *Source* Cradle to
Cradle®

specific business or industrial sector itself. Industrial (technical) or biological goods can be specifically designed or even redesigned to retain its high quality for multiple uses. After the end of life cycle, products could be transferred into high-quality biological or technical nutrients to serve as a high-quality 'raw materials' input for next biological or technical cycle. All materials used by Bauwerk product's life cycle meet the highest standards. For retaining such a high-quality level, products are frequently tested by accredited independent laboratories in more countries. Bauwerk Company is constantly researching to increase its technical quality, in the direction of healthier indoor environment, ecological fairness, and social quality.

4 Certification Program: Cradle to Cradle Certified™

The Cradle to Cradle products Innovation Institute (Cradle to Cradle 2017) is a non-profit organization, administers the sustainable development, and supervises certification process with Cradle to Cradle Certified Product Standard. This organization was specially established to create a new drastic change in production that rearranges the manufacturing and producing of goods into an encouraging strength for environment, public, humans, economy, and the Earth (Fig. 3).

The quality improvements are infinitive and developed the non-profit company by architect William McDonough and Professor Dr. Michael Braungart after more than 25 years of experience of research and growth. The Institute with headquarter in Oakland, California, USA, is directed by a self-governing board of managers. Foundations and individuals who share common values for care of the environment and hope for the future supported non-profit and thus independent Institute's work.

Initial financial funding has been done by the Nationale Postcode Loterij and the Schmidt Family Foundation, founded by Wendy and Eric Schmidt. Remaining self-sustaining revenues will result from certification, licensing fees, and specific training programs (Cradle to Cradle 2017) (Table 1).

Based on the revolutionary book 'Cradle to Cradle: Remaking the Way We Make Things' by William McDonough and Michael Braungart (2002), the Cradle to Cradle

Table 1 ABC—*X* categorization.

IDENTIFYING THE BEST MATERIALS:
ABC-X CATEGORISATION
▬

Goal: Best quality of raw materials, chemicals and ingredients

Category	Description
A	The material is ideal from a Cradle to Cradle perspective for the product in question.
B	The material supports largely Cradle to Cradle objectives for the product.
C	Moderately problematic properties of the material in terms of quality from a Cradle to Cradle perspective are traced back to the ingredient. The material is still acceptable for use.
X	Highly problematic properties of the material in terms of quality from a Cradle to Cradle perspective are traced back to the ingredient. The optimization of the product requires phasing out this ingredient or material.
GREY	This material cannot be fully assessed due to either lack of complete ingredient formulation, or lack of toxicological information for one or more ingredients.
Banned	BANNED FOR USE IN CERTIFIED PRODUCTS This material contains one or more substances from the Banned list and cannot be used in a certified product.

Source Cradle to Cradle®

Certified Product Standard leads inventors, researchers, developers, and producers through a repetitive development procedure (step by step), from the bottom (BASIC level) to the top (PLATINUM level). Cradle to Cradle Certified™ is a product certification, a quality brand and not an eco-label, as it covers all products in all industries within one certification scheme and is not based on limit values for chemicals; however, it is designed to include positively defined ingredients.

Quality is by classification defined into five criteria categories and one recommendation:

- Material Health—MH in short (stage 5.0),
- Material Reutilization—MR (stage 6.0),
- Renewable Energy and Carbon Management—RE&CM (stage 7.0),
- Water Stewardship—WS (stage 8.0),
- Social Fairness—SF (stage 9.0), and
- Recommendations for Product Optimization—RPO (stage 10.0).

5   Material Health

According to Cradle to Cradle, principles are goods with raw materials and all ingredients are assessed through the whole delivery process and precisely weighed for influence on man, animals, and the whole environment. The criteria at each level build towards the expectation of excluding all toxic and unidentified chemicals and becoming nutrients for a safe, environmental friendly continuous cycle.

5.1 Healthy and Safe Materials

The assessment of Material Health generates material substances review ratings evaluation based on the hazards of substances in products and their relative routes of exposure during the intended, and highly likely unintended, use and end-of-use product phases. All ingredients above 100 parts per million (>100 ppm or >0.01%) in a product need to be precisely and critically scientifically assessed according to the Material Health principles. The crucial aim for all goods is manufacturing processes which are only using those substances that are improved and are not prohibited as X or gray evaluated substances.

Naturally, substances are able to reach gradually greater ranks on accreditation level as the parts of substances are improved and environmental friendly raw and recycled materials in the finished product increase.

5.2 Testing Volatile Organic Compound (VOC) Emissions and Results

The goods which are intended for inside use, or influence on living or working inside air pollution, should be strictly in compliance according to the Cradle to Cradle Certified™ VOC emissions prescriptions that are essential for all applications to the Gold and Platinum level of certification and for Externally Managed Components (EMCs) at all accreditation stages. The purpose of this obligation is to guarantee that VOCs are not released from goods that influence the content of VOCs in the inside living or working space, what was confirmed by excellent testing results:

- Volatile Organic Compounds (VOCs) that are checked to be carcinogenic, endocrine disruptors, mutagenic, reproductive toxins, or teratogenic should be below detection limits (i.e., less than 9.0 $\mu g/m^3$ for formaldehyde and less than 2.0 $\mu g/m^3$ for all other chemicals compounds).
- Total Volatile Organic Compounds (TVOCs) must be less than 0.50 mg/m^3.
- Individual VOCs would receive an x assessment criterion, but total contents must be below 0.001 x [TLV/MAK] in any case.
- The period of measurements takes place for at least of 7 days.
- The analytical laboratory that tested and measured values must be ISO 17025 accredited.

6 Material Reutilization

Eliminating the concept of 'waste' is an important goal of sustainable development according to Cradle to Cradle® development concept as a material design exclusively to encourage the conception of an adjusted life cycles of ingredients that totally remove the model of 'waste', just opposite as in the conventional products. The intention is to create encouragements for producers to remove the above-mentioned model of 'waste' by planning innovative goods with substances that may be infinitively cycled. This program encourages producers to be more responsible for proper designing of production systems. Naturally, to achieve a goal in certification process, the all necessities at all lower scores are to be met as well. The further text described the definitions of recommended procedures for new design, research, and development.

6.1 Material Reutilization Score

Standard Requirement according to Cradle to Cradle® principles regulates the next Material Reutilization Score that is obligatory for certain assessment:

- BASIC level: any substance in the material must be declared to fall in a biological or technical cycle.
- BRONZE level: a good has whole life cycle a Material Reutilization Score higher or equal to 35%.
- SILVER level: the good gets whole Material Reutilization Score higher or equal to 50%.
- GOLD level: the good gets a Material Reutilization Score that is higher or equal to 65%. The producer has accomplished a suitable methodology for the material.
- PLATINUM level: the good gets a Material Reutilization Score of exactly 100%. The good's substances are sourced from totally re-covered substances after the use period and reused.

7 Renewable Energy and Carbon Management

Each certification step leads in the direction of carbon-neutral influence and encouraging all activities with total (100%) renewable energy content.

Eco-effective energy production according to Cradle to Cradle® principles follows direction in which production processes in industries and market positively influence the energy supply, ecosystem balance, environment and society. The Renewable Energy and Carbon Management certification level is a mixture of both essential values of sustainable design and development under Cradle to Cradle® system: to generate and use friendly sourced energy and at the same time totally remove the conception of waste.

Renewable energy relocates energy generated from fossil fuels as a consequence is emittance of carbon. Changing the quantity and as well the quality of energy used, influence the balance of CO_2 level in the air and ultimately the climate. Preferably, all emittance could be removed, and energetically needs are generated in surplus to be delivered for nearby society's needs. In the case emissions arise, they are managed as biological nutrients (i.e., without harmful influence to environment).

- BASIC level: producer follows strict quantification of annual electricity use and greenhouse gas emissions, which are associated with the finishing manufacturing production.
- BRONZE level: a sustainable electricity use plan and CO_2 accounting policy are established.
- SILVER level: at end of production level of the goods, minimum 5% of energy have renewable resources or offset with renewable electricity developments, and more than 5% of all GHG emissions are compensated.
- GOLD level: till the end of production of the goods, at least 50% of energy is suitable origin or compensated with electricity developments, and more than 50% of GHG emissions are compensated.
- PLATINUM level: for whole production of the goods, exactly 100% of energy is suitable origin or 100% compensated with suitable electricity developments, and exactly 100% of GHG emittance is compensated. The gray energy related with the production from Cradle to Gate is categorized and measured, besides that is developed a strategy for optimization. At reuse, development on the optimally prepared plan is applicate. More than 5% of the gray electricity connected with the production according to Cradle to Gate principles are balanced by compensation or other sources (e.g., product redesign, proper contractors, reserves in the use stage).

8 Water Stewardship

Those methods are planned to respect water as a valuable and irreplaceable natural reserve for all living beings on our planet. With respect on all levels, that developments are necessary in view of cleaning up effluent, even to drinking water criteria.

8.1 Treating Clean Water as a Valuable Resource and Connection with Fundamental Human Right

Water stewardship generates responsiveness and leads to the water management as an irreplaceable good by stimulating operative controlling and usage policies. Each of the production facility has a significant obligation to maintain this for living nature vital reserve, i.e., water.

- BASIC level: the producer has not reached any destruction of water pollution during 2 years. Water-linked problems are considered; e.g., the producer regulates if shortage is a problem, and subtle environment and ecology environments are at danger due to straight procedures. Water stewardship program cares that achievement is reached, with demonstrations of re-application, and significant development on action procedures.

- BRONZE level: water audit is finalized on a production site.
- SILVER level: (i) chemicals as a result of production processes in water effluent are detected and evaluated, or, (ii) supply connected to water subjects (a suitable influence plan is established).
- GOLD level: (i) optimization of all chemicals in effluents, which are related to production processes (chemicals recognized as problematic….), or, (ii) Silver-level requirements requested a development on the strategy, what is demonstrated (production sites with no problematic effluents).
- PLATINUM level: drinking water quality standards are strictly required for all water leaving the manufacturing productions.

9   Social Fairness

All activities inside companies are planned to protect all humans and environment in such a way that development could be established and positive influence on the humankind and the environment is ensured.

9.1   Positive Support for Social Systems

Social Fairness stage in certification assessment confirms that development is reached in direction to sustainable operations that bring to all owners an interest, also to employees, society participants, customers, and the environment. For business, ethics is necessary to overcome the boundaries of the existing company habits.

- BASIC level: an inside operation audit is made to confirm existence of basic human rights. Organization processes confirm those topics that are established. For re-application is required a demonstration of progress on the management plan.
- BRONZE level: a self-audit of full social responsibility is completed; besides, positive impact strategies, which established on documents 'UN Global Compact Tool' or 'B-Corp.', are developed.
- SILVER level: one of the next issues is closed: (i) raw substances content to a minimum of 25% of the raw substances by weight is finished successfully, or, (ii) fully investigated supply side connected to minimal social requirements, and a suitable influence plan is developed, or, (iii) the producer is guiding an active advanced society program which stimulatingly influences employees' lives, the local community, global community.
- GOLD level: two requirements of the Silver score are finished.

- PLATINUM level: a third party completed production site audit according to a worldwide accepted social responsibility platform, as 'SA8000 standard' or 'B-Corp'; besides, all Silver-level prescriptions are finished.

10   The Case Study Bauwerk's 'Silente' Parquet

Operating, manufacturing, or assembling activities with wood means taking care and responsibility of Nature and environment. Wood is one of the most important re-growing natural raw substance in practically all regions and is assumed to be the building substance of the future. Illegal deforestation and depletion reduce this issue for forthcoming generations. For the wood productions, it is essential to take responsibility for sustainable forestry and wood supply, which is fundamental and vital from an ecological standpoint. This is necessary and important for ecology, environment, and whole Nature. Exclusive wood supply from ecological forestry was always operation standard for Bauwerk Company. Furthermore, Bauwerk Company always confirmed to use of advanced solvent-free adhesives and lacquers, implementing the extensive system of recycling in the production procedures; the last innovative techniques are used to lower dangerous emissions and high employees and the environment safety prescriptions, all according to ISO 14001 standards.

The strict execution of guidelines by supplying natural wood materials and final marketable goods is a significant producer's methodology. For all Bauwerk wood products as well as parquet products, wood raw materials are sourced from sustainable forestry.

By adapting a design and a new innovation under the name of "Silente", Bauwerk is breaking new ground. Silente products follow the sustainable development under the principle of Cradle to Cradle® design, which means a closed raw material cycle and a respective quality assurance production process.

10.1   Bauwerk Parkett AG Company Profile

Bauwerk Parkett is a Swiss company with a long-standing tradition and combines reliability and precision in the manufacture of top-quality parquet and other wooden floors. Proud of its origins, the company constantly strives toward innovation. In the last few years, Bauwerk has seen significant development and has set out on a new course, converting from a technological and production-oriented enterprise to a market-oriented company with a focus on healthy living, sustainability, and design.

Over two-thirds of Bauwerk products are manufactured in St. Margrethen (Switzerland). Bauwerk Parkett AG has two further factories in Kietaviskes (Lithuania; since 2014) and Ðurđevac (Croatia; since 2017). Bauwerk's product portfolio comprises 350 articles, ranging from two-layers to three-layer as well as solid parquet. Bauwerk sold approximately 4.1 million square meters of parquet in 2016.

10.2   Parquet for Generations

Bauwerk Parkett (2017) is designed in such an innovative way that substances constantly stayed in closed loop life cycle and no part of the product ever changed into waste material or final product need excessive non-renewable energy. Those substances in product stayed in product life cycle even for more centuries; all needed water inputs are preserved and cared; and company behaves in a fair and socially responsible manner in relation to its employees as well as to the public and society. On this successful way, accreditation assistance from EPEA Switzerland company is helping Bauwerk in application of the Cradle to Cradle® program, its sustainable design and development of products, and as general assessor supporting for its goal to reach the Cradle to Cradle Certified™ successful authorization.

All ingredients, wood and other input substances supplied by 36 delivering companies, are carefully checked with Material Health certification steps, Material Reutilization score value, impact on environment and checked in whole life cycle loop into reusable substance original state. Bauwerk's manufacturing facilities as well as all suppliers were subjected to close scrutiny during the new 'Silente' technology certification process.

The parquets and other wooden floors that include the innovative 'Silente' know-how could simply be taken up without any effort (thanks to the new 'Silent-Mat' innovative solution), and what is very important is it could be anytime returned to the company. Naturally, all returned components are either recycled or re-conditioned for new quality products. Because of this infinitive close cycle, Bauwerk conserves the valuable wood resources and acts in the interest of environment and future generations. The products that include the 'Silente' innovative know-how are produced completely from substances that are harmless for both the environment as well as for humans.

10.3   Innovative Story of 'Silente' Products Was Awarded with Gold Certification Level

Bauwerk Parkett Company is always fastened with glue totally to the below surface of construction, which guarantees the improvement that a perfect connection is secured to the below surface, for ideal insulation against transmission of impact noise. Besides that is very important that skilled workers are trained in practical and theoretical level to assure an ideal assembling of the wooden floor product.

The sustainable design and continuous development were confirmed by Cradle to Cradle Certified™ Gold-level prize for Bauwerk's 'Silente' wooden floor elements, because:

- the strict consumption of health and thus non-hazard substances,
- the all raw and recycled substances are always a need,

- the valuation of environmental friendly energy sources and maintenance of carbon management,
- environmentally compatible water management and water stewardship program,
- social responsibility to workers and society,
- Bauwerk Parkett is ISO 14001 certified.

10.4 Circular Economy—Task, Innovation, and Implementation

Process of the disassembling of an attached conventional wooden or parquet floor is hard and not friendly activity; besides that the removal work loud, unclean, and exhausting, it also takes a lot of time- and patience-consuming and thus costly. Additionally, the wooden floor is destroyed during removal and must be discarded. Because of the different material components, reuse of the such waste is impossible.

It is more and more important from professional as well as from user site to find an efficient innovation connected with insulation against walking and impact sound noise. These needs followed by starting a new program in 2011, when Bauwerk Company with helpful support of the German manufacturer company WPT designed and developed an innovative walking and impact noise insulating mat. The goal was to reach strong innovative know-how, high technical acceptance, and ideal environment friendliness. The solution as a result of mutual research between both companies was finished by use of 80% content of environmental friendly substances (chalk and polyurethane glue), positioned just below and above the mat, with a function similar to existing solutions. This new and innovative mat together with the application method is named as 'Silent-Mat,' was patented by producer, and is exclusively made only for Bauwerk Parkett company in manufacturing format as 7.5 m × 1.0 m in rolls. The montage starts by gluing of innovative mat on the floor surface, followed by direct gluing of every Bauwerk Parkett or other wooden floor product.

During time and experience with new innovations was a new solution of wooden floor product for material health, their use, recovering and reuse was developed, by maintaining its high value over its entire lifetime and multiple lifecycles. The idea which came up was to develop the parquet floor nondestructively, so the wooden floor can be used repeatedly in a closed loop and multiple product cycles. Even more, those new know-how solutions are successfully combined into one solution; by direct montage in the factory of new innovative mat rolls on the bottom side on finished surface of parquet, and later on the construction site of the floor.

EPEA Switzerland (2017) is supporting and supervising productions with various products and activities in different sustainable research and development process for application of Cradle to Cradle® certification, which is run with an experienced with plenty of practice cases, educated, and multidisciplinary functioning team. In the Alpine region, parts of Cradle to Cradle® assessors are implemented in all productions. Chemical reviews on scientific level and other chemical evaluations for all certification developments are generated in tight collaboration with EPEA Internationale Umweltforschung GmbH, Hamburg, Germany.

Parquet Cleverpark is available in the dimensions 1250 mm × 100 mm. The wooden floor system was successfully tested for impact sound insulation at the EMPA institute in Dübendorf, Switzerland.

Around 1,500 m^2, 'Cleverpark Silente' has been applicated for a major project in Chur, Switzerland, what was at the beginning major 'Silente' applied new innovative wooden floor area and thus presented the testing surface in view of effective acoustical insulation properties. All this was reached with use of innovative solution 'Cleverpark Silente,' and the relation to the results obtained in laboratory and values in the certificates was confirmed. The scientific institute EPEA Switzerland assessed in 2013 the Silver certification level according to Cradle to Cradle Certified™, and the chemical optimization was done in 2015 for Gold level.

The Bauwerk Company and their products have been evaluated in five main certification criteria and one recommendation:

- Material Health—MH,
- Material Reutilization—MR,
- Renewable Energy and Carbon Management—RE&CM,
- Water Stewardship—WS,
- Social Fairness—SF, and
- Recommendations for Product Optimization—RPO.

11 Luxury

Sustainable products practically share the same values of qualities and other criteria of luxury goods. Sustainable luxury products inherit extraordinary creativity and design, outstanding materials with good quality, and properties such as durability. The context «less but better» shares the spirit of sustainable luxury. Luxury goods should have implicit sustainability built in; they are long-life products and consequently do not go out of fashion.

Sustainable luxury products offer the consumer fulfillment of all their individual values and desires by introducing material healthier consumer choices rather than making them feel guilty.

Luxury products have not been promoted as sustainable. The «green» touch was not desired, but today products with «invisible 2 beauty» in design with the right materials and chemicals used are recognized from consumers. Leading brands are taking steps in this trend and are presenting sustainable luxury as an essential part of their own brand image. In addition, social and environmental awarenesses are increasing and global resources come under investigations; this is creating a new environment for the luxury industry.

Bauwerk created a clear pronounced marketing message to the consumer (Bauwerk Parkett 2017).

Why is this so important? We are spending during living or working periods from minimum 80 up to 90 or even longer part of our live periods inside space, needing daily from 10 to 20 m^3 of fresh air. Beside that airtight building walls, roofs, and windows, what represent building envelope, make sense in terms of energy, ventilation, so, to reduce harmful substances. Appropriate construction and finishing elements are an unconditional need for reaching a guarantee for non-sick inside air. Bauwerk Company has been testing quality and environmental friendliness of their products since year 2010 together with the Sentinel Haus

Institute.

Is parquet healthy? No. Even though wooden substances in construction products are a precious bio-based and environmental friendly product, the producers of wooden floors implement additional sticking substances, protecting layers, and different finishing. The highest standards and regular testing procedures are essential for all materials used by Bauwerk Company independently of the local country production.

11.1 Innovative Sound Reduction

Our ears perceive intrusive sounds or fast changes in sound levels as noise. Glued parquet substantially reduces footfall sounds. Bauwerk's 'Silente' technology is an innovative system solution that reduces footfall and impact sounds even further, an effect that is particularly sought-after in renovation projects.

11.2 Invisible but Measurable

High environmental and material health standards in closed living or working rooms are not successfully developed by coincidence; proper design, well-oriented project, labor checking, and increasing technical and environmental values. Separately from environmental friendly inside room air and proper room acoustics, Bauwerk's 'Silente' innovative wooden floor solution solved important strong positive arguments: absence of electrical static charge; it feels much warmer than any other floor surface, also visual, and beside that natural wood surface, it is temperature-friendly in summer (enough cool) as well as in winter (friendly warm).

Odors in buildings, especially those that occupants perceive as disturbing, may have many different sources; among them are building materials such as paints, varnishes, wooden materials, glues and insulation materials used during the construction or renovation of buildings. Fixtures and fittings, for instance, furniture or office equipment, affect the emergence of odors. Bauwerk Parkett has no smell, and rooms can be used immediately after its installation.

Table 2 Respective threshold values are substantially lower than those required

Parameter	DIBt/AgBB2015[a]	Eco-Institute label
TVOC after 3 days	10.0 mg/m^3	3.0 mg/m^3 (=3000.0 μg/m^3)
TVOC after 28 days	1.0 mg/m^3	0.30 mg/m^3 (=300.0 μg/m^3)
Formaldehyde	0.120 mg/m^3	0.0360 mg/m^3 (=36.0 μg/m^3)
After 28 days	0.120 mg/m^3	0.0360 mg/m^3 (=36.0 μg/m^3)
Ammonia	0.10 mg/m^3 (=149.0 μg/m^3)	100.0 μg/m^3
Individual substances—reference values for interior environment	No	Yes
Material test	No	Yes
Odor test	No	Yes
Phthalate softening agents	Generally allowed	Generally prohibited
Halogen compounds	No specific requirements	Generally prohibited

Source Bauwerk
[a]Switzerland is following the European standards

Bauwerk Parkett flooring has been tested by independent, qualified institutes to determine its health properties and has successfully passed the strict Eco-Institute criteria, including those applying to building assessments by the Sentinel Haus Institute, DGNB, or Leed. The respective threshold values are substantially lower and more extensive than those required for general building authority approval or quality labels for other floor coverings, as can be seen in the Table 2.

11.3 Healthy Living—Invisible but Measurable

We intensively care for humans as well as for whole Nature, because we do not inherit the Earth from our Ancestors; we borrow it from our Children.

Everybody who works with natural bio-based materials has a tender intuitive relation for natural origin of certain product and has special relation to environment, energy use, and whole life cycle loop. That is why Bauwerk Parkett Company develops continuous relationship to healthy living and working inside room environment, what is included in the next properties:

- the special, fashionable wood and other raw substances, totally recyclable and are produced exclusively in environmental friendly way,
- to avoid all toxic or other environmental unfriendly substances in whole tightly closed life cycle loop, i.e., all wooden substances, with proper production, to product and to living or working indoor environment,
- open and fair partnerships with final users, customers, dealers, montage groups, all suppliers, and their numerous subcontractors,

- to be responsible and very reliable partner to whom you can rely on,
- to manufacture exclusively market leading wooden surfaces in quality and environment-friendly aspects, acoustically pleasant, reliable and environmental and to nature acceptable way.

12 Silente Technology

'Innovation and Leadership are the only survival strategies' emphasized Kälin (2017) of EPEA Switzerland and longtime expert on environment, certification, and textile. The management of Bauwerk Parkett AG created with a visionary mindset, profound management skills, and courageous decisions to generate a recognized lighthouse position of a company implementing sustainable development with the Cradle to Cradle® design with a holistic approach throughout the company, as:

- visionary mindset to create a solution to safe resources for coming generations,
- moral imagination 'transforming nature' (Gorman 1998),
- redesign of adhesive system for entire company,
- implementing innovation and knowledge management strategies,
- integrating 'closed loop concepts' into business,
- changing of the business model in taking goods back,
- product responsibility for coming generations,
- product lifetime extension for three generations from 25 to 75 years,
- marketing approach 'healthy living,'
- integration into the entire supply chain, customers, consumers and take back concepts.

Bauwerk's 'Silente' technology is an innovative system solution that reduces footfall and impact sounds even further, an effect that is particularly sought-after in renovation projects. 'Symbiosis between man and nature' described all operations within responsible economy. People who work with wood as a raw input material develop through production process a special feeling and relation to our Nature, everything related on the 'old but innovative' theory of a symbiosis between humankind and Nature (Fig. 4).

As a result, the system called 'Cleverpark Silente' gets for the topic of Material Health Silver level and in the next four topics: Material Reutilization, Renewable Energy and Carbon Management, Water Stewardship, Social Fairness the Gold level, everything according to principles of Cradle to Cradle Certified™. The assessment covered all materials and their chemical substances above 100 parts per million level in regard to their environmental and health relevance criteria, such as; mutagenicity, reproductive toxicity, carcinogenicity, teratogenicity, endocrine disrupting activity, biodegradability/persistence, toxic poisonousness, substances with possibility of irritating of skin surfaces or eyes, and aquatic harmfulness. The raw substances for the parquet floor are precisely tested for the allowed level of toxic substances and the level of organ halogens (bromine, chlorine and fluorine). The cost of the whole designing

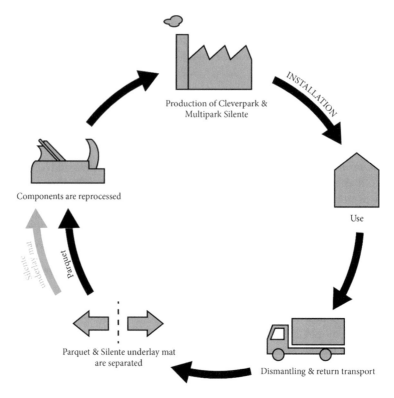

Fig. 4 So-called 'closed loop' is the principle of 'Silente' technology for wooden floors. *Source* Bauwerk

and developing primary group of five employees and additionally external consulting team was high. All chemical components and raw materials were evaluated at EPEA International Umweltforschung in Hamburg.

The certificate level Silver for five issues for Parquet has been awarded in October 2013, whereas in January 2015, the new higher Gold-level project Cradle to Cradle Certified™ has been started. Furthermore, the suppliers were requested for upgrade certification and thus replaced the 'Gold-critical' materials due to optimization process, because a criteria Material Health reached higher Silver level and in the rest of four other chapters reached the Gold status.

Without an active collaboration, this could not be achieved and consequently Bauwerk's assortments 'Clever Park Silente' and 'Multi-Park Silente' have been since June 2015 for the sustainable development confirmed with Cradle to Cradle Certified™ Gold-level certificate, which is unique and the only producer with healthy living wood parquet. All achievements, generated knowledge, and results from the above-mentioned C2C® project was successfully brought in entire Bauwerk Swiss Parquet productions, so now the all Bauwerk Parkett production in Switzerland are according to Cradle to Cradle Certified™ standards at Gold or Bronze.

Fig. 5 'Silente parquet'.
Source Bauwerk

Bauwerk Company with its products is committed to have finished a tremendous and important step in direction to sustainable environment. The goal in the near future is to use less valuable timber for the same requirements. An important contribution to sustainability is made through the multiple uses of the products.

13 Conclusions

The sustainable development according to Cradle to Cradle® (2017) program increases the profitability in the whole life cycle closed loop of a certain material. All problems in the delivery or supply, the industrial processes and implementation, accomplish much better position. The economy itself, the eco-related indirect costs, and the social parts are thus much more expectable and become profitable.

All needed ingredients and raw materials through the whole supply process are actuality during the process of sustainable development considered from the beginning as the raw wooden substances to final materials in the concept Cradle to Cradle® design. Therefore, the above-mentioned resulted into products of incomparable quality. Finally, an infinitive use of material substances is now accomplished without any limitations (Fig. 5).

References

Arkar, C., Domjan, S., & Medved, S. (2018). Lightweight composite timber façade wall with improved thermal response. *Sustainable Cities and Society*. https://doi.org/10.1016/j.scs.2018. 01.011.

Bauwerk Parkett (2017). Healthy living brochure. Retrieved in Nov 15, 2017 from http://www. bauwerk-parkett.com/de.html.

Braungart, M. (1992). An intelligent product system to replace waste management Braungart, Engelfried. *Fresnius Envir Bull, 1*, 613–619. Basel, Switzerland: Birkhauser Verlag. 1018-4619/92/090613-07S1.50-0.20/0.

Cradle to Cradle Certified^{TM} Trademark and Cradle to Cradle® C2C® Copyright (2017). Registered trademarks of McDonough Braungart Design Chemistry (MBDC). Cradle to Cradle Certified^{TM} is a certification mark licensed exclusively by the Cradle to Cradle Products Innovation Institute (C2CPII), all rights reserved. Retrieved on Nov 15, 2017 from www.c2ccertified.org.

Dovjak, M., Shukuya, M., & Krainer, A. (2012). Exergy analysis of conventional and low exergy systems for heating and cooling of near zero energy buildings. *Journal of Mechanical Engineering, 58*(7/8), 453–461.

EN 15804 (2012). Sustainability of construction works—Environment product declarations—Core rules for the product category of construction products, European Standard. European Committee for Standardisation.

Gilbert, N. L., Guay, M., Gauvin, D., Dietz, R. N., Chan, C. C., & Levesque, B. (2008). Air change rate and concentration of formaldehyde in residential indoor air. *Atmospheric Environment, 42*(10), 2424–2428.

Gilding, P. (2011). *The great disruption* (p. 16). New York: Bloomsbury Press.

Gorman, M. (1998). *Transforming nature. Ethics, invention and discovery*. ISBN 0-7923-8120-3.

Gustavsson, L., Pingoud, K., Sathre, R. (2006). Carbon dioxide balance of wood substitution: Comparing concrete- and wood-framed buildings, Springer, Greenhouse gas balances in building construction: Wood versus concrete from life-cycle and forest land-use perspectives, Environmental and Energy Systems Studies, Lund University, Lund Institute of Technology, Lund, Sweden, Energy Policy.

Hawken, P., Lovins, L. H. (1999). *Natural capitalism, creating the next industrial revolution*. Little, Brown and Company. ISBN 0-316-35316-7.

Heimlich J. E. (2008). Formaldehyde. The invisible environment fact sheet series. Retrieved on Nov 15, 2017 from http://ohioline.osu.edu/cd-fact/pdf/0198.pdf.

ISO 14025 (2009). Environmental labels and declarations—Type III environmental declarations—Principles and procedures, standard. Geneva, Switzerland: International Standards Organisation.

ISO 14040 (2006). Environmental management—Life cycle assessment—Requirements and guidelines, standard. Geneva, Switzerland: International Standards Organization.

ISO 14067 (2013). Carbon footprints of products, standard. Geneva, Switzerland: International Standards Organization.

Jelle, B. P. (2011). Traditional, state-of-the-art and future thermal building insulation materials and solutions—Prosperities, requirements and possibilities. *Energy and Building, Elsevier,*. https:// doi.org/10.1016/j.enbuild.2011.05.015.

Kälin, A. (2017). Interview with Albin Kälin. Oct 10, 2017.

Kim, C. W., Song, J. S., Ahu, Y. S., Park, S. H., Park, J. W., Noh, J. H., et al. (2001). Occupational asthma due to formaldehyde. *Yonsei Medical Journal, 42*(4), 440–445.

Kitek Kuzman, M., Kutnar, A. (2014). Contemporary Slovenian timber architecture for sustainbility, green energy and technology. Switzerland: Springer. ISSN 1865-3529.

Košir, M., Krainer, A., Dovjak, M., Perdan, R., & Kristl, Ž. (2010). Alternative to conventional heating and cooling systems in public buildings. *Journal of Mechanical Engineering, 56*(9), 575–583.

Kunič, R. (2017). Carbon footprint of thermal insulation materials in building envelopes. *Energy Efficiency*. https://doi.org/10.1007/s12053-017-9536-1.

Kutnar, A., Hill, C. (2014). Assessment of carbon footprinting in the wood industry. In: S. S. Muthu (Ed.), *Assessment of carbon footprint in different industrial sectors*, vol 2 (Eco-Production). Singapore [etc.], pp. 135–172. Berlin: Springer.

Lakrafli, H., Tahiri, S., El Houssaini, S., & Bouhria, M. (2017). Effect of thermal insulation using leather and carpentry wastes on thermal comfort and energy consumption in a residential building. *Energy Efficiency*. https://doi.org/10.1007/s12053-017-9513-8.

McDonough, W. (1993). Essay: A centennial sermon: Design ecology ethics and the making of things.

McDonough, W., & Braungart, M. (2002). *Cradle to Cradle, remaking the way we make things*. New York: North Point Press.

Paganin, G., Angelotti, A., Ducoli, C., et al. (2017). Energy performance of an exhibition hall in a life cycle perspective: embodied energy, operational energy and retrofit strategies. *Energy Efficiency*, pp 1–22. https://doi.org/10.1007/s12053-017-9521-8.

Pajek, L., & Košir, M. (2017). Can building energy performance be predicted by a bioclimatic potential analysis? Case study of the Alpine-Adriatic region. *Energy and Buildings, 139,* 160–173.

EPEA Switzerland (2017). EPEA Switzerland GmgH. Retrieved on Nov 15, 2017 from www.epeaswitzerland.com.

Ansgar Igelbrink, 1964 has completed Diploma in Business Administration, and from 1990 onwards, he has worked within the building distribution segment until 2000, when he changed to a segment leading chemical company for wooden flooring. Since 2007, he has developed and led the most successful brand Bauwerk Parkett within the wooden flooring industry. One of the success factors was the focus on healthy living, which was trustworthy documented with the C2C certification in 2013.

Albin Kälin, 1957, Merchant on textiles was awarded in 2001 with UBS Key Trophy as the 'Rhine Valley Entrepreneur of the Year.' From 1981 to 2004, he was Managing Director of Rohner Textile AG in Switzerland. Under his leadership, the company won, since the 1990s, 19 international design awards. In 1993, he stimulated the development of the product line Climatex® (https://www.climatex.com) and thus the first Cradle to Cradle® products worldwide. From 2005 to 2009, he worked as CEO of EPEA Internationale Umweltforschung GmbH in Hamburg. In 2007, he was additional CEO of EPEA Netherland. From 2009 to the present-day, Albin Kälin is founder and CEO of EPEA Switzerland GmbH.

Marko Krajner, 1972 starts his career in 1999 as wood science and technology engineer in furniture company Gorenje in Slovenia. Later he was quality manager, project manager, purchasing manager in different companies in Slovenia, in foundry, automotive, boating, wood and machining sectors. In 2011, he found 3ZEN consulting company based on Lean production and administration, TRIZ principles, 6-sigma and other continuous improvement and systematic innovation tools. He is Lead auditor for wood based certification schemes FSC since 2011 and PEFC since 2013. In 2017, he became General accredited assessor for Cradle to Cradle Certified™ certification Behalf of EPEA Switzerland. From the starting of 2018 to present-day, Marko Krajner is founder and CEO of 3ZEN d.o.o.

Roman Kunič is an associate professor, researcher and a head of Chair of buildings and constructional complexes. He finished his bachelor's degree in 1986 and in 1990 a master's degree, both at UL FGG. In 1997, he finished MBA on Clemson University in South Carolina, USA. In 2007, he finished his Ph.D. thesis 'Planning an assessment of the impact of accelerated ageing of bituminous sheets on constructional complexes.' After 25 years of experience in the economy, mainly related to the industry of building insulation materials, he returned to the faculty. R & D work: new thermal, sound and waterproof insulations, advanced materials (e.g., VIP and PCM), environmental and sustainable assessment, analysis of accelerated ageing and lifetime prediction, and analysis of dynamic thermal response of buildings. He is an innovator of five patent applications. He works as a reviewer in 'Gradbeni vestnik', Energy Efficiency, Journal of RMZ—materials and geoenvironment. Also works as a lecturer in the areas of building envelope design, advanced materials, building renovation and sustainable building assessment. He is also mentor or co-mentor on more than 80 theses. He holds membership in professional and scientific associations: IZS—Slovenian Chamber of Engineers, ISES—International Solar Energy Society, 'Čar lesa' 'The Magic of wood' Ljubljana, a member of the organizing committee.

Selected Bibliography

ZRIM, Grega, MIHELČIČ, Mohor, SLEMENIK PERŠE, Lidija, OREL, Boris, SIMONČIČ, Barbara, KUNIČ, Roman. Light distribution in air-supported pneumatic structures : comparison of experimental and computer calculated daylight factors. Building and environment, 2017, 108, 1–43. https://doi.org/10.1016/j.buildenv.2017.04.005 [COBISS.SI-ID 8052321].

KUNIČ, Roman, OREL, Boris, KRAINER, Aleš. An Assessment of the Impact of Accelerated Ageing on the Service Life of Bituminous Waterproofing Sheets. Journal of materials in civil engineering, 2011, 23, 1746–1754. https://doi.org/10.1061/(ASCE)MT.1943-5533.0000326. [COBISS.SI-ID 5509985].

KUNIČ, Roman, MIHELČIČ, Mohor, OREL, Boris, SLEMENIK PERŠE, Lidija, BIZJAK, Aleš, KOVAČ, Janez, BRUNOLD, Stefan. Life expectancy prediction and application properties of novel polyurethane based thickness sensitive and thickness insensitive spectrally selective paint-coatings for solar absorbers. Solar energy materials and solar cells, 2011, 95, 2965–2975. https://doi.org/10.1016/j.solmat.2011.05.014. [COBISS.SI-ID 5509729].

KUNIČ, Roman, KOŽELJ, Matjaž, OREL, Boris, SURCA, Angelja Kjara, VILČNIK, Aljaž, SLEMENIK PERŠE, Lidija, MERLINI, Dušan, BRUNOLD, Stefan. Adhesion and thermal stability of thickness insesitive spectrally selective (TISS) polyurethane-based paint coatings on copper substrates. Solar energy materials and solar cells, 2009, 93, 630–640. [COBISS.SI-ID 4117018].

Trends of Sustainable Development Among Luxury Industry

Jitong Li and Karen K. Leonas

Abstract The luxury sector is a well-established global industry worth approximately US$200 billion a year; although successful, the luxury market is quietly being re-framed to align with key and emerging trends in the industry. The concept of sustainability is gaining increased attention by the industry and consumers. Many companies including H&M, Levi's, and Nike have already incorporated sustainable development and supply chain partner that focus on sustainability into their business models. Sustainability being inconsistent with value associated with the luxury sector, which is a leading sector known for high margins and social reputation, reports have criticized luxury brands for lagging behind others with regard to sustainable development. Considering their 'value network' in a conventional way and ignoring emerging needs and social conditions may be the reasons why the existing luxury companies have not adopted sustainability practices more quickly. Under increasing pressure to implement sustainable development throughout the industry, some new luxury entrepreneurs are emerging with remarkable perspectives on sustainable development. They break the traditional business innovation known to the luxury sector and are implementing the concept of sustainable development as a direction in their business strategies. In addition, they are moving toward developing a circular economy to realize 'sustainability' in their supply chains. In this chapter, redefinition of luxury, trends in the luxury market, adoption of sustainability among luxury brands and consumers, disruptive business model innovation, and the circular economy are discussed. At last, a case study on sustainable luxury swimwear entrepreneurs is presented.

Keywords Sustainable development · Luxury · Entrepreneurs · Luxury swimwear · Disruptive business strategy · Circular economy

J. Li (✉) · K. K. Leonas
Textile and Apparel, Technology and Management, North Carolina State University, Raleigh, NC, USA
e-mail: jli62@ncsu.edu

K. K. Leonas
e-mail: kleonas@ncsu.edu

© Springer Nature Singapore Pte Ltd. 2019
M. A. Gardetti and S. S. Muthu (eds.), *Sustainable Luxury*,
Environmental Footprints and Eco-design of Products and Processes,
https://doi.org/10.1007/978-981-13-0623-5_6

1 Introduction

The term 'luxury' is always followed by exclusive, expensive, beautiful, attractive, and pleasant; reversely, the term 'sustainability' is related to environment, social equity, and economics. These two terms seem irrelevant, but recently they are being combined, and a new term 'sustainable luxury' has appeared. The sustainable development in the luxury sector is the outcome of both pressures and motivation.

For the existing well-known luxury brands, stress pressures from society, criticisms, other industries, and consumers' increasing concerns regarding sustainability are pushing them forward into sustainable development. The existing luxury brands are making efforts to embrace sustainable development, whereas they keep looking back concerned with the conflict between the nature and characteristics of sustainability versus luxury. Some of them are afraid that the differences between 'luxury' and 'sustainability' will be the reason of failing sustainability practices. They are also worried that their long-time established reputation might be damaged. Others are challenged by the difficulties in sustainable development as they have high volumes, existing product positioning, and production lines. It is difficult for them to embrace a radical change to sustainable development and production.

Considering these factors, some new luxury brands are arising emerging with excellent perspectives and practices with regard to sustainable development. These companies are motivated to implement sustainable luxury production practices in response to consumer and industry concerns. They make bold attempts for the aim of building radical sustainable development. The traditional luxury strategy is broken, and the sustainable business strategy is used as a direction of their development. In addition, circular economy is embraced to supervise and control the sustainable development process. In consideration of all the efforts made by these emerging companies and the risks taken by them, they are regarded as entrepreneurs.

Both existing luxury brands and emerging luxury entrepreneurs are making changes in response to sustainable development, but what are the differences and which one is more competitive? These questions are emerging and need to be examined to better understand the trends of sustainable development in the luxury sector. To answer the questions, the luxury swimwear industry and three emerging sustainable luxury swimwear entrepreneurs were selected for study in this article and to explaining sustainable development practices in the luxury sector. This chapter begins with a brief overview of sustainable development in the luxury sector, followed by a comparison of luxury and sustainability product characteristics and a review of business model innovation and circular economy within this sector. A case study on emerging sustainable luxury swimwear entrepreneurs is presented to support the discussion.

2 Background

2.1 *Sustainable Development Pressure on the Luxury Sector*

The luxury sector is changing from a conventional to a more sustainable paradigm with the pressure from within, from their consumers, and from other industries. Today, an increased number of consumers are expressing concerns regarding the sustainability of products within the textile, apparel, and fashion industries. They are demanding products that are less toxic, more durable, and made from recycled materials (Lozano et al. 2010). Companies within the fashion, textile, apparel, and retail (FTAR) supply chain are also focusing on increasing their sustainable efforts through the implementation of policies focusing on fair treatment of employees and reducing their environmental impact. With the increasing attention on the concept of sustainability from diverse industries and consumers throughout the world, sustainable development is regarded as a global topic (Law and Gunasekaran 2012). Nevertheless, sustainable products account for less than 1% of the total market share (The Co-operative Bank 2009; Davies et al. 2012), indicating sustainability is more limited in application than expected or the literature sometimes suggests. Reports highlighting the growth of sustainable consumption tend to focus on low-value, commoditized product categories such as food-related products, cosmetics, and apparel (Davies et al. 2012). These categories have also seen the highest market shares in terms of the sale of sustainable products (The Co-operative Bank 2009; Davies et al. 2012). As one of the categories with successful sustainable development, today, many fashion and apparel companies, including H&M, Uniqlo, NIKE, The North Face, Patagonia, and New Balance, recognize the importance of sustainability in business and therefore have incorporated it into their supply chain.

Under the context of thriving sustainable development in other non-luxury industries throughout the world, criticisms from consumers on the luxury sector have returned. Historically, the luxury sector was condemned in planet, people, and profit, although not all the '3Ps' were regarded as the meaning of sustainable development at that time. In the 1980s and 1990s, there were the anti-fur campaigns; many brands and luxury retailers have since eliminated the use of fur in their products or taken measures to ensure animal welfare conditions in their fur supply chains. In the 1990s, numerous sweatshop scandals, including extremely low wages, long working hours, and use of child labor, were brought to the attention of the public and pressure was put on companies to implement factory-compliance monitoring programs to prevent these injustices. Even though criticisms were somehow resolved during that period, today, the pendulum has swung back, some of the previous criticisms have returned, and new criticisms have emerged. This is in some cases due to increased awareness of these violations as a result of globalization of the industry, better connectedness through technological advancements, and the prevalence of social media. After 2000, more than 100 designers, including Vera Wang, John Galliano and Dolce and Gabbana, are showing fur in their collections. Fashion arbiters such as Vogue magazine are putting fur on the cover again (Ramchandani and Coste-Maniere 2017). In addition, dissatisfaction from the society about environmental destruction appeared, and in the past several years, the luxury sector has faced intensifying criticism about footprint (Gazzola et al. 2017).

The luxury sector of the apparel and textile industry is extremely sensitive to reputational damage; consumers in all social classes are increasingly concerned about social and environmental issues and prefer sustainable products that reflect their own values and beliefs (Bendell and Kleanthous 2007; Hennigs et al. 2013). With this in mind, the concept of sustainability has to become a priority for luxury brands as well. The luxury sector is currently establishing the concept of 'sustainable luxury.' The efforts of the luxury sector in moving forward with sustainability-related programs can be shown from the sustainable products launched by luxury companies, sustainable manufacturing, and cooperation between luxury companies and sustainability-related organizations.

Although some factors, such as the sustainable development pressure from other industries and increasing consumers' demands on sustainability related luxury products, are contributing to moving the luxury sector forward to sustainable development. However, in the luxury sector, very few established brands have moved to a primarily sustainable model as there is perceived risk. Existing studies provide evidence that the luxury industry is perceived by experts and consumers to lag behind other industries with regard to sustainable development (Bendell and Kleanthous 2007; Hennigs et al. 2013). Many elements can be considered as reasons for the failure of this sector to embrace sustainability through their supply chain. One of the most prominent reasons is the difference in the nature and characteristics between luxury and sustainability.

2.2 Emerging Sustainable Luxury Entrepreneurs

Under this complicated and difficult situation, some new luxury brands are emerging that aim to solve issues related to sustainable development and make benefits. These new brands are being designed, developed, and promoted by entrepreneurs. J. C. Mill was the first to utilize the term 'entrepreneur' among economists. Direction, supervision, control, and risk taking are considered as the functions of the entrepreneurs (Brockhaus and Horwitz 1986). Correspondingly, these emerging luxury brands are making clear direction, advisable supervision and control, and overcoming risk associated with sustainable development. In contrast with incumbent well-known luxury brands, these entrepreneurs producing smaller volumes (compared with existing brands) are able to more easily target their products, designs, and brands to be sustainable. They are using 'sustainable development' as their business model strategy, which has not been used before in the luxury sector. In addition, for the aim of circular economy development, they are also implementing sustainable development in their product lines.

This chapter introduces sustainable development in the luxury sector, compares sustainable development in existing well-known luxury brands and luxury

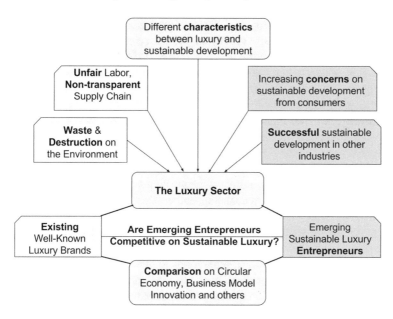

Fig. 1 Comparison between luxury brands—existing and emerging (*Source* Made by the authors)

entrepreneurs from business model innovation and circular economy perspectives, and evaluates whether there are opportunities for entrepreneurs in the area of sustainable development. Characteristics of luxury and sustainability, business model innovation, circular economy, and a case study on sustainable luxury swimwear entrepreneurs are presented here. Figure 1 shows the model of comparison between luxury brands—existing and emerging; it is also the framework of this article. The left part of this figure (yellow colored boxes) shows facts about the current luxury sector and existing well-known luxury brands. The right side of the figure (red colored boxes) shows additional current facts and changes related to sustainable development in the luxury sector and emerging luxury entrepreneurs. The basic structure of this article is explained by the middle section (represented by the blue boxes). Here, the chapter will begin at the top of the figure with the inherent conflict between sustainable products/practices and items thought of as luxury goods.

3 Luxury Versus Sustainability

Luxury can be particularly difficult to define, since a strong element of human involvement, very limited supply, and the recognition of value by others are key components (Cornell 2002). According to Kapferer (1997), the word luxury 'defines beauty; it is art applied to functional items. Like light, luxury is enlightening. Luxury items provide extra pleasure and flatter all senses at once.' Although the term 'luxury' elicits a variety of different descriptions, this article will use the luxury definition of McKinsey (1990) as 'whose price and quality ratios are the highest in the market' to make 'luxury' easier to be understood and the later research on emerging luxury swimwear entrepreneurs realizable. Whereas, it is difficult to distinguish a luxury product from non-luxury product just with the definition of luxury. A deeper understanding and additional characteristics of luxury have been identified by Wiedmann et al.'s research (2009) on dimensions of luxury value, which include price value, usability value, quality value, uniqueness value, self-identity value, hedonic value, materialistic value, conspicuousness value, and prestige value in social networks. Corresponding with the statement of Sjostrom et al. (2016), these dimensions reflect some attributes of luxury, including premium price, premium quality and craftsmanship, authenticity, brand architecture, exclusive and limited, pedigree and heritage, method of production, and recognizable styles.

According to the World Commission on Environment and Development (1987), sustainability is defined as 'development that meets the needs of the present without compromising the ability of future generations to meet their needs.' This definition emphasized natural resource protection and rational development and use of natural resource. It is a logical definition for that period, since the first scientific contributions on sustainability focused on the use of natural resource and their influence on quality of life at that time (Robinson 2004). However, with the development in sustainability, it has been recognized that sustainability involves complex and changing dynamics. Further and deeper understandings of sustainable development have been developed which rely on the intersection of three important principles related to the environment, social equity, and economics. The social principle requires that everyone is treated fairly and equitably. The economic principle requires the adequate production of resources so that society can maintain a reasonable standard of living, and the environmental principle asserts that society protects its environmental resources (Bansal 2002). These principles can be transformed at the company level as 'respect for people (at all levels of the organization), the community, and its supply chain; respect for the planet, recognition that resources are finite; and generating profits that arise from adhering to these principles' (Joy et al. 2012).

Table 1 shows a comparison between sustainable and luxury product characteristics based on the descriptions provided. By comparing the components of sustainable development with those of luxury goods, it becomes possible to identify potential elements of contradiction. The first five characteristics listed show that the nature of a luxury good is opposed to those related of sustainability, since luxury is related to ostentation and margin, but sustainability is related to the fairness or social harmony facets. However, the following four characteristics (six to nine) are in agreement for sustainable products and luxury goods. One of the fundamental principles of a true luxury strategy is to be produced locally by talented artisans and to respect the sources of raw materials (Kapferer 2010). This is also one approach sustainable development uses to realize sustainable production. If luxury brands continue to follow this principle, then luxury goods that are also considered sustainable should be more prevalent today. Unfortunately, the rapid growth of the luxury goods sector has attracted new brands focused on volume and increased profit margins; they deviate from the strict

Table 1 Comparison between sustainable and luxury product

Characteristic	Sustainable product	Luxury product
1. Society	Fair and equitable	Self-identified; exclusive
2. Environment	Protection	Not in concern
3. Economic benefits	Sustainability first	Margin first
4. Price	Wide range	Expensive
5. Supply chain	Transparent	Less transparent
6. Raw materials	Eco-friendly	Respect source of raw materials
7. Production method	Sustainable technology/handmade	Handcraft and others
8. Consumer	Sustainability concerned	With various purposes and higher income
9. Value	Eco-friendly, harmony social, determined by consumer (self-lifestyle and of lifestyles of others in the community)	Attitudinal, contextual, and personal values as well as a habit or a routine

Note The information in this table is based on the work of Kapferer (2010), Kapferer, and Michaut (2015) with contributions of the authors to reflect luxury and sustainable product information

luxury rules to increase their margins and move away from its fundamental principles (Kapferer and Michaut 2015). The value of a luxury product is also based on the consumer's perspective; hence, if the value of sustainability can be regarded as a value of luxury by consumers, the connection between sustainability and luxury will be developed. The textile and apparel sector of the luxury industry is under pressure to develop more sustainable products. The 'sustainability' is being monitored by consumers and social media. If companies that produce luxury products continue to only focus on ways to increase the profits and volume, they will come under criticism and negative reviews from consumers. However, until consumers begin to make their purchase decisions based on sustainability as well as luxury, these companies will not promote or pursue the development of sustainable luxury. The more consistent the values between sustainable and luxury products are, the more likelihood there will be for sustainable development.

4 Disruptive Innovation—Business Model Innovation

As discussed earlier, the value of both luxury and sustainability is based on consumer's perspective, and at the same time, concerns from consumers on sustainable luxury have increased. To satisfy the new requirement, the luxury sector is implementing sustainable development in different ways. Some of them use the concept of sustainability as an innovation to build add-value strategy; others regard sustainability as a disruptive innovation and direction of their business strategy through all their development phases.

4.1 Conceptual Work

There are two specific types of disruptive innovations—(1) business model innovations and (2) radical product innovations which are also known as disruptive technology (Markides 2006). At first, just product innovations, like superior technologies, were considered as disruptive innovations. Over time, Christensen widened the application of the term to include not only new and superior technologies but also business models (Markides 2006). Business model innovation which is also called strategic innovation is the discovery of a fundamentally different business model in an existing business (Markides 2006). Disruptive strategic innovations share certain characteristics: First, they emphasize different product or service attributes; second, disruptive strategic innovations usually start out as small- and low-margin businesses; Third, disruptive strategic innovations grow to capture a large share of the established market (Charitou and Markides 2002).

Based on the nature of luxury and sustainability, and their inherent conflict in specific properties, (Table 1), there is need to reconcile some of the conflict. Enhancing the 'sustainability' of a luxury product will add benefits for luxury brands (Tynan et al. 2010), including attracting new customer segments, preserving social values, and protecting the environment. Sustainability is a new concept to luxury brands; therefore, it can be said that sustainable luxury is a disruptive innovation. Since it is not a kind of or a set of certain products (or disruptive technology), but a concept (disruptive strategic innovation) but rather a change in the business model—or a business model innovation. It is promoting a new lifestyle (eco-friendly), and it is changing patterns of consumption, production, or exchange for a positive societal outcome.

Historically, luxury brands have collaborated with technology development companies searching for the technology for sustainable products. For example, on the Web site of luxury brand Stella McCartney, there is the following statement identifying the company's commitment to sustainability. 'Sustainability is evident throughout all her collections and is part of the brand's ethos to be a responsible, honest, and modern company' (About Stella 2017). An example one program that addresses producing more sustainable products uses a regenerated cashmere yarn, Re.Verso™, to replace the traditional cashmere. Re.Verso™ is made in Italy using a new system developed through the collaboration of several organizations and companies including Green Line and Nuova Fratelli Boretti (Re-Verso 2017). Green Line is responsible for the sourcing and sorting of pre-consumer textile waste from all of Italy and select European countries. Nuova Fratelli Boretti is responsible for the rigorous hand-picked selection and the mechanical transformation process that produces high-quality woolen fibers. This example supports sustainability being a disruptive strategic innovation.

Disruptive strategic innovations frequently begin with small businesses or entrepreneurs rather than in established companies. There are trade-offs required when innovations are implemented, and many times, it is difficult for an established company to respond to the different or new innovation efficiently and effectively. A company that attempts to compete in both arenas (luxury and sustainability) simultaneously risks degrading the value of its existing activities and, as a result, is likely to experience major inefficiencies. Attempts to manage the innovation by utilizing the company's old systems, processes, incentives, and mind-sets will hinder and possibly destroy the new business (Charitou and Markides 2002). The characteristics of disruptive strategic innovations are suitable for the innovation of sustainable luxury; hence, there are more opportunities for entrepreneurs to venture into the development and marketing of sustainable luxury and be successful.

4.2 Respond to Disruptive Strategy Innovation

According to the work of Charitou and Markides (2002) and Chen and Miller (1994), two factors that influence if a company responds to major disruptions in their businesses are (1) do they have the ability to respond and (2) do they have the motivation to respond. For an established company, the ability to respond is determined by several factors including the company's portfolio of skills, its resources, and the time it has at its disposal. But most important is the nature and size of the conflicts between the traditional business and the new business. The higher the degree of conflict, the lower is their ability to respond (Charitou and Markides 2002). In terms of sustainable luxury, the conflict between the nature of luxury and the nature of sustainability is strong. As a result, there is difficulty for established companies to adapt sustainable luxury practices. On the other hand, the company's motivation to respond is determined by factors such as the rate at which the innovation is growing and how threatening it is to the main business. So, the more strategically related the new business is to existing strategies, the more motivated the company will be to respond. There are many differences among established luxury companies regarding the relationship between their original business model and a sustainable luxury business model. Therefore, it can be difficult to identify the degree of motivation at the levels of all established luxury companies. Overall, it may be a struggle for existing luxury brands to analyze these two factors, ability and motivation; hence, it will also be complex for them to make responding decisions. Whereas, for emerging entrepreneurs, they still need to consider these factors, their business model may embrace. It is easier for entrepreneurs to make a decision, since this will be their first or second innovation without influencing their inexistent initial strategies.

The business model innovation of sustainable luxury has been shown to be implemented in two different ways. For existing luxury brands, it is a value-added strategy. The prestige of a luxury brand may be reinforced, and the exclusivity of the brand is increased. Sustainability in this case is regarded as an additional attribute to the pre-existing luxury product. Many famous brands, such as Gucci and Hermès, moved in this direction (Gazzola et al. 2017). However, for some emerging luxury brands, sustainability may be conceived of as an original source of luxury. In this type of experience, the sustainable resource for luxury does not increase the perceived value of a pre-existing product, but generates an exclusive property. This latter aspect appears the true and ultimate business model innovation of sustainable luxury. This can also be examined by looking at actual examples. Some Web sites of established well-known luxury brands are searched randomly from Table 2, like Dior (handbags made from bio farms), Chanel (cooperates with local handcraft artisans), and Coach (less CO_2 emission). All of them have sustainability-related projects or products (those can be found on the Web site of Kering Group), but there is not 'sustainability' following their brand names, even though there is not 'sustainability' (or related terms) on the homepage. In contrast, the Web sites of the emerging entrepreneurial companies (Table 2) were examined for information related to sustainability and they all included some information. For example, Tengri (knitwear with sustainable Khangai Nobel Fibers), Everest Isles (sustainable swim shorts) and Tortoise Denim (sustainable wiser wash), 'sustainability' (and related terms) was easy to find along with additional information on sustainable luxury. These entrepreneurs are using the concept of sustainable luxury as a business model throughout their supply chain, from product design to retailing. While established famous brands (identified in Table 2) are adopting sustainability in some products and supply chain practices, it is not a founding principle of the brand; there are more risks associated with a major change for an established brand which makes adoption more difficult.

Overall, luxury entrepreneurs are incorporating sustainability through the supply chain and capitalizing on the benefits discussed here. They are taking advantage of applying sustainability and promoting it as advantage of their product. The approaches they are using to survive with this innovation and the benefits they can obtain will be discussed later.

5   Circular Economy

After adopting sustainable development as a business model innovation, the luxury sector is searching for approaches to realize sustainable luxury. One of the approaches they are embracing is a circular economy as opposed to the more traditional linear economy. The concept of a circular economy is not new and was given a theoretical foundation in the field of industrial ecology in the early 1990s, following the concept of sustainability (Ayres and Simonis 1994; Bocken et al. 2016). Circular economy (CE), also known as a 'closed-loop' economy, is an industrial and social evolutionary concept that pursues holistic sustainability goals through a culture of no waste (De los Rios and Charnley 2017). It is described as a scientific development model, where resources become products, and the products are designed in such a way that they can be fully recycled (Yap 2005). It can be regarded as a practical approach or model implementing the innovation of sustainability. The circular economy model contrasts with the traditional linear business model of production of take-make-use-

Table 2 Luxury brands and sustainable practice

Category	Brand name	Sustainable practice
Incumbent luxury brands	Dior[a]	Handbags made from biofarms
	Chanel	Cooperates with local handcraft artisans
	Coach	Less CO_2 emission
	Bottega Veneta (Kering 2017)	Shoes made with Caiman Sandal, protecting regional species diversity
	Vivienne Westwood (Garcia 2011)	Launched a collection of upcycled bags and iPad cases in partnership with the UN and the World Trade Organization
	Gucci (Kering 2017)	With the Italian fashion house releases the world's first handbag collection made from sustainable Amazonian leather back in 2013
Emerging luxury entrepreneurs	Tengri (2017)	Listed in Sustainia100 as one of the world's leading sustainable business solutions, produces Khangai Noble Fibres® via a 100% transparent supply chain
	Everest Isles (2017)	'Uses bluesign® approved textiles. The bluesign® system is the solution for a sustainable textile production. It eliminates harmful substances right from the beginning of the manufacturing process and sets and controls standards for an environmentally friendly and safe production'
	Tortoise Denim (2017)	'Our method uses natural and biodegradable additives along with ozone, eliminating the need for corrosive chemicals while using little to no water to achieve that true vintage look'
	Mara Hoffman (2017)	Econyl®, polyester: recycled polymers, cotton: Global Organic Textile Standard, Lenzing Tencel, modal or rayon made from the pulp of sustainably harvested trees. Working in partnership with artisans in India in an effort to create sustainable employment that is socially and environmentally responsible
	Kilometre Paris (2017)	Hand embroidered linens from Kilometre Paris encourages the fashion world to buy linen instead of cotton (indigenous, less water to produce)
	EDUN (2017)	In respect of its mission to source production and encourage trade in Africa, EDUN mixes its modern designer vision with the richness and positivity of this fast-growing continent. EDUN is building long-term, sustainable growth opportunities by supporting manufacturers, community-based initiatives and partnering with African artists and artisans'

[a]Citation for Dior, Chanel, and Coach is the same: (Kapferer 2010). It can be found in the reference list

dispose and an industrial system largely reliant on fossil fuels, because the aim of the business shifts from generating profits from selling artifacts to generating profits from the flow of materials and products over time (Bakker et al. 2014; Bocken et al. 2016). The potential and benefits of circular economy are huge. At the global level, a CE could help enable developing countries to industrialize and developed countries to increase well-being and reduce vulnerability to resource price shocks. For companies, it offers a model of sustainable growth fit for a world of high and volatile resource prices (Preston 2012).

In reviewing existing and emerging luxury companies and their sustainability-related operations, it can be concluded that there are two directions luxury companies are implementing to embrace circular economy. Most of the sustainable practices of existing luxury companies are related to product design, whereas the practice of emerging sustainable luxury companies is focused on a business strategy (coincide with the earlier discussion on disruptive sustainable innovation of existing and emerging luxury brands). Based on this, circular economy is introduced from two perspectives, product design strategy and business model strategy. There are three approaches to achieve CE: (1) slowing, (2) closing, and (3) narrowing resource loops (Bocken et al. 2016). Interestingly, these are also the three characteristics that distinguish the circular economy from a linear economy. Both CE product design and CE business model strategies should correspond to one of these three approaches.

CE Product Design Strategy

According to the work of Bocken et al. (2016), there are design strategies for slowing resource loops (including designing long-life products and design for product-life extension) and for closing resource loops (including design for a technological cycle, design for a biological cycle, and design for dis- and reassembly). In consideration of the specialty of luxury industry, high quality and durability, not all the product design strategies from the work of Bocken can be applied to promote CE or sustainable development. Currently, the most frequently used strategy is 'design for a technological cycle.' Design for a technological cycle aims to develop products in such a way that the materials can be continuously and safely recycled into new materials or products, for example, Gucci's sustainable leather and Stella McCartney's regenerated cashmere.

CE Business Model Strategy

Several CE business model strategies are identified in the work of Bocken et al. (2016) and Baker et al. (2014). Under business model strategies for slowing resource loops, there are access and performance models, extending product value, classic long-life models, and encouraging sufficiency. Under business model strategies for closing loops, there are extending resource value and industrial symbiosis. Like CE product design, not all the strategies are suitable for sustainable luxury, and some of the strategies can be combined. For example, Eviana Hartman, a New York-based designer and founder of Bodkin, a sustainable fashion line established in 2008, had the goal of merging a particular aesthetic and a focus on sustainability (5 Impressive Sustainable Luxury Fashion Brands 2017). The company uses organic and recycled fabrics and non-harmful dye in the production of the apparel. Hartman noted that the demand for organic cotton has increased, and currently, the demand is greater than the supply, which means the end-product will be more expensive. Harman also believes that beautiful designs made with attention to detail by skilled people will always cost more than mass-manufactured clothing. She included both 'extending product value' and 'extending resource value' in her business model.

Although CE was introduced over 25 years ago, and it has had increasing interest from various companies and governments, 'circular economy business models have not yet made the world a better place (Planning 2015).' There are reasons which are manifold and partly rooted in conceptual flaws of our world economic order as well as in the inherent irrationality of consumer behavior (Planning 2015). Achieving a closed-loop economy is more difficult than expected, especially for the luxury sectors. More efforts need to be put into extracting benefits from CE.

6 Case Study—Emerging Sustainable Luxury Swimwear Entrepreneurs

Some comparisons between existing luxury brands and emerging sustainable luxury entrepreneurs have been discussed, including business model innovation and circular economy. The purpose of the case study here is to provide giving deeper and practical introduce on the emerging entrepreneurs. Before the analysis of three entrepreneurs in the luxury swimwear industry, some basic research on the luxury swimwear industry in general was completed to have a review on the swimwear industry and determine which is a sustainable luxury swimwear entrepreneurship. Considering different culture and respects to swimwear in different regions (e.g., some of the un-affluent Asian countries remain reluctant to allow females, either adults or children, to wear swimwear for leisure, as a fashion statement or for sporting exercise), just the North American swimwear market was researched in this work.

The North American (NA) retail swimwear market is relatively active. It reflects the culture where both sexes, adults, and children are expected to wear swimwear for leisure and for sporting exercise. Research showed that the women's market dominated the NA swimwear market accounting for 71% of the value total. The men's share was 16%, girl's 10%, and boys just 3%. Table 3 shows the average retail price of swimwear by category for both the North American market and world (including North America) for 2009. The data shows that there is a pronounced price imbalance between the categories with the women's swimwear being much higher priced than the other categories. This retail price for this category is also well above the average for both NA and worldwide markets. According to Newbery (2015), women are prepared to spend far more per piece for fashion, while men focus more on utility. In the case study presented here, only women's and men's swimwear brands were included. In addition, it is important to note that not all well-known luxury brands retail adult swimwear every season (i.e., Gucci); or they just have very

Table 3 2009 retail swimwear average prices per retail piece (US$)

Region	Total	Women	Men	Girls	Boys
North America	$18.58	$27.79	$13.06	$10.69	$4.50
World	$12.36	$18.78	$8.14	$6.72	$4.32

Source Newbery (2015)

Table 4 Luxury brands swimwear price range (US$)

Brand name	Women			Men
	Bikini-top	Bikini-bottom	One-piece	Trunks/shorts
Burberry[a]	N/A[b]	N/A	N/A	$85–295
Dolce & Gabbana	$225–325	$225–255	$495–695	$195–475
Stella McCartney	$140–215	$100–190	$210–425	N/A
ERES	$250–595		$270–715	N/A
On the Island	N/A	N/A	$415–765	N/A
Michael Kors	N/A	N/A	$358–698	N/A

[a]Price source of these six brands in Table 4 is Saks Fifth Avenue (2017). Citation can be found in the reference list
[b]In these tables of this article, N/A means the price cannot be found

few swimwear pieces on sale (i.e., Chanel and Louis Vuitton). As a result, in this study, prices of luxury swimwear were collected from luxury online stores or online information from stores that carry luxury brands (i.e., Saks Fifth Avenue 2017). A characteristic of luxury products is their higher price when compared with the average market price. Table 4 shows the swimwear price ranges by brand, by gender, and by style, and Table 5 shows the swimwear price ranges of the entrepreneurs selected. Comparing the price of women's swimwear in Table 3 (North American average) and 5 (Emerging Luxury Entrepreneur Brands), the luxury entrepreneur brands are about three times higher than average price of swimwear. Comparing the prices of men swimwear for the same groups (NA retail prices with Luxury Entrepreneurs Brands) of Everest Isles ranges from four to nineteen times higher than average price of non-luxury swim shorts. Comparing data in Table 4 with that in Table 5, it shows that although these luxury entrepreneurships do not have the price as high as the price of the established luxury brands with a long history (Table 4), their prices can fall into the luxury price range. These three entrepreneurial companies conform to the price characteristics of luxury swimwear.

Another characteristic of luxury products is high quality. All these entrepreneurial brands use diverse techniques and/or fabrics to ensure high quality. This will be analyzed as a component of their innovation and business models for the three selected brands, Mara Hoffman, Jeux De Vagues, and Everest Isles.

Table 5 Emerging luxury entrepreneurs swimwear price range (US$)

Brand name	Women			Men
	Bikini-top	Bikini-bottom	One-piece	Trunks/shorts
Vitamin A (2017)	$79–130	$78–120	$146–192	N/A
Mara Hoffman (2017)	$95–155	$95–125	$195–250	N/A
Manakai Swimwear (2017)	$96–98	$96–98	$198	N/A
Jeux De Vagues (2017)	$116–129	$112	$226–239	N/A
Everest Isles (2017)	N/A	N/A	N/A	$74–255

6.1 Mara Hoffman

Mara Hoffman founded her label in 2000 and currently serves as President and Creative Director of the privately owned New York City-based company. The lifestyle brand produces women's ready-to-wear and swimwear and is known for its signature prints. Hoffman's goal is to foster mindful consumption habits and encourages consumers to reevaluate the relationship society has with clothing. Although this company was not founded on principles of sustainability, they soon began to focus on using more sustainable materials and that was followed by changes in their production methods working internationally to develop a more sustainable supplier base. Hoffman is a successful transition case from a general luxury brand to a sustainability-driven luxury brand.

Hoffman and her team pay close attention to how they improve from production design through production. In 2017, she was recognized with the Positive Impact Award for 'Brand Leadership in Advancing Sustainability'. The sustainable operations of Mara Hoffman can be categorized through their materials, manufacturing, and cooperation as described below.

Materials: All Mara Hoffman swimwear is made from recycled fibers, either PET, made from post-consumer recycled polymers to reduce waste going into landfills, or Econyl® 100% regenerated nylon fiber made from fishnets and other nylon waste. In their most recent Spring 2018 collection, all of the cotton used is certified through GOTS (Global Organic Textile Standards). They also introduced hemp, one of the more sustainable renewable fibers. Growing hemp does not require irrigation or pesticides and actually replenishes soil quality, making it a great companion crop for farmers.

Manufacturing: The fabrics are printed digitally which reduces water and chemical use and waste. They prioritize and work with dyes and chemical companies that are eco-conscious. In addition, they are transitioning into compostable and recyclable packaging in everything from poly bags to stickers to boxes to our hang tags.

Cooperation: In May 2017, Mara Hoffman formed a new partnership with Nest, a nonprofit that facilitates connections between high fashion brands and artisan cooperatives. Mara Hoffman has worked with a Delhi, India-based embroidery and beading group for several years, and in conjunction with Nest developed a plan for training and developing to help to expand and diversify this Indian supplier base and to enable its business to become a more sustainable model. The brand is also a member of the Sustainable Apparel Coalition (SAC), an industry-wide international coalition of brands, suppliers, and retailers who are working together to implement a standardized reporting system to improve the social and environmental impact of the global fashion industry. (DuFault 2017; Mara Hoffman 2017)

Although Mara Hoffman was not initially formed as a sustainability innovative luxury brand, as a smaller brand, with a shorter history (less than 20 years), it has successfully adopted a business model that aligns with sustainable development. They have embraced many components of the circular economy through changing its raw materials to regenerated nylon and certified cotton; adopting more sustainable manufacturing practices; and developing a relationship with international and local governments to establish a more sustainable supplier base focusing on both the environmental and social aspects. Mara Hoffman is working to build an integrated closed-loop system for her products. It will be more sustainable and influential other sustainable luxury entrepreneurs.

6.2   Jeux De Vagues

Jeux De Vagues was created by Katherine Terrell in 2015. It is a brand with smaller volumes and focuses only on women's swimwear. Their sustainability operations can be categorized through their approach to materials, fair labor, and cooperation with organizations (Jeux De Vagues 2017).

Materials: Jeux De Vagues uses recycled fabrics in their products. The solid fabrics are made with Econyl®, a 100% regenerated nylon fiber made from pre- and post-consumer materials. Their printed fabrics are made from recycled polyester produced from recycled water bottles. In addition to using more sustainable fabrics, they use fully biodegradable hangtags that quickly dissolve in water.

Fair Labor: Their bikinis are produced in a fair labor factory in Los Angeles that pays fair wages under fair labor practices regulated by the state of California.

Cooperation: They are committed to giving back to the environment and are a member of 1% For the Planet (an international organization whose members contribute at least one percent of their annual sales to environmental causes).

As a sustainability innovation driven entrepreneur, this brand knows exactly, they are using recycled nylon as our base material is by no means the perfect solution to the environmental problems, but it is a place to begin. They hope that with every Jeux De Vagues bikini their customers wear, a conversation with others about what individuals can do for the health of oceans can begin. This 'hope' is helping to reinforce the self-identify value of luxury products and brands' differentiation. As a new brand in its second year, there are only about 11 items on its Web site and it is still too early to tell the benefits or loss.

6.3   Everest Isles

Established in 2012, Everest Isles creates modern essentials for life in and around the water. Designed and made in the USA, Everest Isles uses globally sourced luxury fabrics to develop technical men's swimwear and nautical inspired sportswear (Everest Isles 2017). Everest Isles is a pioneer in the area of men's sustainable luxury swim trunks. They are implementing sustainability into their fabrics and manufacturing primarily through the use of more sustainable fabrics and using sustainably certified manufacturing processes…

Materials: Everest Isles uses fabrics made from Econyl® to reduce the ocean pollution. Nylon is also used as a raw material in of swimwear, and if nylon-made products, like fish nets, are discarded into oceans, they will end up there and wreak havoc on marine animals and the environment (Lusher et al. 2014). Whereas, Econyl® is a regeneration system can be endlessly repeated without any loss in material quality (Econyl 2017). Nylon waste is collected through different initiatives and projects and then regenerated; in this way, ocean pollution is reduced.

Manufacturing: Everest Isles uses bluesign® approved textiles. The bluesign® certification is the possible solution for a sustainable textile production. It eliminates harmful substances from the beginning of the manufacturing process and sets and controls standards for an environmentally friendly and safe production. There are five principles in this system, resource productivity, consumer safety, water emission, air emission, and occupational health and safety (Bluesign 2017).

Although a new company, Everest Isles, has been identified as a luxury brand due to the price point, unique fabrics, high-quality zippers, special Marlow cordages (Everest Isles 2017), and their products are placed in shops with high visibility including YOOX and M5 SHOP. They are getting benefits, like positions in luxury market with little competition due to few brands focusing on men's sustainable luxury swimwear, from the innovation of sustainable luxury, and further development will be seen.

7   Discussion and Conclusions

In this chapter, various aspects of sustainable development have been introduced, discussed, and compared between existing well-known luxury brands and emerging entrepreneurs in the luxury sector. Table 6 is a summary of the comparisons. The comparisons in this table consist of three categories: brand competitiveness, sustainable products, and sustainable development. By comparing aspects under each category, the conclusion can be obtained.

Table 6 Comparison between entrepreneurs and existing brands on SD in the sector of luxury

Category	Sub-category	Existing brands	Emerging entrepreneurs
Brand competitiveness	Company scale	Larger	Smaller
	Brand reputation	Higher, stronger	Lower, less strong
	Social influence	Strong	Weak
	Retail platform	Multiple	Limited
Sustainable products	Product species ratio	Lower	Higher
	Product price	Higher	Lower
Sustainable development	Business model	Add-value strategy	Radical and fundamental strategy
	Circular economy	Less done	More done
	Consumers	More existed consumers	More emerging consumers

Firstly, about the brand competitiveness, existing brands have better performance than emerging entrepreneurs. That is the result of existing brands' long history and abundant resource on human, social, and market. Although it seems to have nothing to do with sustainability, brand competitiveness can build a powerful foundation for sustainable development, including advertising, retailing, and educating consumers on sustainability. Emerging entrepreneurs are weak on brand competitiveness, but they are stronger on their sustainable products. They have higher sustainable species ratio (sustainable product species/all product species) and lower price. Lastly, the comparison is on the sustainable development. Emerging entrepreneurs are implementing sustainable development in a deeper, broader, and more direct way. They are breaking the traditional paradigm in the luxury sector, and as a result, their target market is somehow different. The existing brands are persuading their existing consumers to adopt their sustainable luxury products, while the emerging entrepreneurs are attracting the younger, emerging consumers care about sustainable development.

In general, existing brands have an advantage in the area of brand competitiveness, while emerging entrepreneurs are better on sustainable products and sustainable development. Emerging entrepreneurs can be competitive in the sustainable luxury sector, but it cannot be judged that emerging entrepreneurs are more competitive than existing brands. If they want to survive in the sustainable luxury sector, these entrepreneurs need to fill their gaps. They have more pressure to attract consumers and to build their brand's reputation, but there are opportunities. As discussed earlier, although existing brands have had some sustainable development practices and multiple retail platforms, they do not incorporate many details related to sustainability on their Web sites or elsewhere. Emerging entrepreneurs in this sector who are incorporating sustainability into their business model and provide a great deal of information regarding sustainability on their websites, however they need to make additional efforts to introduce themselves and continue to highlight their focus on sustainability to ensure this is known in the market. Their Web sites have high visibility, and their education to consumers about sustainable luxury will work as expected.

References

5 Impressive Sustainable Luxury Fashion Brands. (2017, February 2). Retrieved from https://theluxauthority.com/sustainable-luxury-fashion-brands/.

About Stella. (2017, November 2). Retrieved from https://www.stellamccartney.com/experience/us/about-stella/.

Ayres, R. U., & Simonis, U. E. (1994). *Industrial metabolism: Restructuring for sustainable development.*

Bakker, C., den Hollander, M., Van Hinte, E., & Zljlstra, Y. (2014). *Products that last: Product design for circular business models.* TU Delft Library.

Bansal, P. (2002). The corporate challenges of sustainable development. *The Academy of Management Executive, 16*(2), 122–131.

Bendell, J., & Kleanthous, A. (2007). *Deeper luxury: Quality and style when the world matters.* WWF-UK.

Bluesign. (2017). Retrieved from https://www.bluesign.com/.

Bocken, N. M., de Pauw, I., Bakker, C., & van der Grinten, B. (2016). Product design and business model strategies for a circular economy. *Journal of Industrial and Production Engineering, 33*(5), 308–320.

Brockhaus, R. H., & Horwitz, P. S. (1986). The psychology of the entrepreneur. *Entrepreneurship: Critical Perspectives on Business and Management, 2,* 260–283.

Charitou, C. D., & Markides, C. C. (2002). Responses to disruptive strategic innovation. *MIT Sloan Management Review, 44*(2), 55–64.

Chen, M. J., & Miller, D. (1994). Competitive attack, retaliation and performance: An expectancy-valence framework. *Strategic Management Journal, 15*(2), 85–102.

Cornell, A. (2002). Cult of luxury: The new opiate of the masses. *Australian Financial Review, 47.*

Davies, I. A., Lee, Z., & Ahonkhai, I. (2012). Do consumers care about ethical-luxury? *Journal of Business Ethics, 106*(1), 37–51.

De los Rios, I. C., & Charnley, F. J. (2017). Skills and capabilities for a sustainable and circular economy: The changing role of design. *Journal of Cleaner Production, 160,* 109–122.

DuFault, A. (2017, October 17). *Positive impact awards: Mara Hoffman wins brand leadership in advancing sustainability.* Retrieved from https://bkaccelerator.com/positive-impact-awards-mara-hoffman-wins-brand-leadership-in-advancing-sustainability/.

Econyl. (2017). Retrieved from http://www.econyl.com/.

EDUN. (2017). Retrieved from https://edun.com/.

Everest Isles. (2017). Retrieved from https://www.everestisles.com/pages/about-everest-isles.

Gazzola, P., Pavione, E., & Pezzetti, R. (2017). Sustainable consumption in the luxury industry: Towards a new paradigm in China's high-end demand. In *Proceedings of the 2nd Czech-China Scientific Conference 2016.* InTech.

Garcia, P. (2011, June). *Need it now: Vivienne Westwood recycled bags.* Vogue. Retrieved from https://www.vogue.com/article/need-it-now-vivienne-westwood-recycled-bags.

Hennigs, N., Wiedmann, K. P., Klarmann, C., & Behrens, S. (2013). Sustainability as part of the luxury essence: Delivering value through social and environmental excellence. *The Journal of Corporate Citizenship, 52,* 25.

Jeux De Vagues. (2017). Retrieved from https://jeuxdevagues.com/.

Joy, A., Sherry, J. F., Jr., Venkatesh, A., Wang, J., & Chan, R. (2012). Fast fashion, sustainability, and the ethical appeal of luxury brands. *Fashion Theory, 16*(3), 273–295.

Kapferer, J.-N. (1997). Managing luxury brands. *Journal of Brand Management, 4,* 251–260.

Kapferer, J. N. (2010). All that glitters is not green: The challenge of sustainable luxury. *European Business Review*, 40–45.

Kapferer, J. N., & Michaut, A. (2015). Luxury and sustainability: A common future? The match depends on how consumers define luxury. *Luxury Research Journal, 1*(1), 3–17.

Kering. (2017). Retrieved from http://www.kering.com/en/sustainability.

Kilometre Paris. (2017). Retrieved from https://kilometre.paris/.

Law, K. M., & Gunasekaran, A. (2012). Sustainability development in high-tech manufacturing firms in Hong Kong: Motivators and readiness. *International Journal of Production Economics, 137*(1), 116–125.

Lozano, J., Blanco, E., & Rey-Maquieira, J. (2010). Can ecolabels survive in the long run?: The role of initial conditions. *Ecological Economics, 69*(12), 2525–2534.

Lusher, A. L., Burke, A., O'Connor, I., & Officer, R. (2014). Microplastic pollution in the Northeast Atlantic Ocean: Validated and opportunistic sampling. *Marine Pollution Bulletin, 88*(1), 325–333.

Manakai Swimwear. (2017). Retrieved from https://manakaiswimwear.com/.

Mara Hoffman—Our Approach. (2017). Retrieved from http://www.marahoffman.com/world-of/about/our-approach/.

Markides, C. (2006). Disruptive innovation: In need of better theory. *Journal of Product Innovation Management, 23*(1), 19–25.

McKinsey. (1990). *The luxury industry: An asset for France*. Paris: McKinsey.

Newbery, M. (2015). *Global market review of swimwear—forecasts to 2019: 2015 edition: Swimwear market history, 2009–2013*. Bromsgrove: Aroq Limited. Retrieved from http://proxying.lib.ncsu.edu/index.php?url=http://search.proquest.com/docview/1656580821?accountid=12725.

Planing, P. (2015). Business model innovation in a circular economy reasons for non-acceptance of circular business models. *Open Journal of Business Model Innovation, 1*, 11.

Preston, F. (2012). *A global redesign?: Shaping the circular economy*. London: Chatham House.

Ramchandani, M., & Coste-Maniere, I. (2017). To Fur or not to fur: Sustainable production and consumption within animal-based luxury and fashion products. In *Textiles and clothing sustainability* (pp. 41–60). Singapore: Springer.

Re-Verso Supply Chain. (2017, October 29). Retrieved from http://www.re-verso.com/en/info/la-supply-chain-integrata.

Robinson, J. (2004). Squaring the circle? Some thoughts on the idea of sustainable development. *Ecological Economics, 48*(4), 369–384.

Saks Fifth Avenue. (2017, November 20). Retrieved from https://www.saksfifthavenue.com/Entry.jsp.

Sjostrom, T., Corsi, A. M., & Lockshin, L. (2016). What characterises luxury products? A study across three product categories. *International Journal of Wine Business Research, 28*(1), 76–95.

Tengri. (2017). Retrieved from http://www.tengri.co.uk/.

The Co-operative Bank. (2009). *The ethical consumerism report 2009*. Retrieved November 11, 2017 from http://www.co-operative.coop/ethicsinaction/sustainabilityreport/Ethical-ConsumerismReport/.

Tortoise Denim. (2017). Retrieved from https://tortoisedenim.com/.

Tynan, C., McKechnie, S., & Chhuon, C. (2010). Co-creating value for luxury brands. *Journal of Business Research, 63*(11), 1156–1163.

Vitamin A. (2017). Retrieved from https://www.vitaminaswim.com/.

Wiedmann, K. P., Hennigs, N., & Siebels, A. (2009). Value-based segmentation of luxury consumption behavior. *Psychology & Marketing, 26*(7), 625–651.

World Commission on Environment and Development. (1987). *Our common future*. Oxford: Oxford University Press.

Yap, N. T. (2005). Towards a circular economy. *Greener Management International*.

A Circular Economy Approach in the Luxury Fashion Industry: A Case Study of Eileen Fisher

Sabine Weber

Abstract Past studies have considered opportunities and barriers to sustainability in the fashion industry, or consumer perspectives and disposal behaviour, but few studies have explored a circular economy approach as a chance for business extension. There is a lack of research that examines luxury fashion, circular economy and entrepreneurship, showing how a circular economy could look. This study used a case study approach to observe and analyse the circular economy business model of Eileen Fisher (EF), New York. This study explores how the company has developed its take-back programme and how this programme led to the development of recycling operations at EF. In 2017, twelve semi-structured interviews were conducted with employees from EF, representing different departments and operating at various functions in the company. Their responses were analysed according to a content analysis method to outline EF's approaches to both luxury fashion and a circular economy additional data from the company was obtained. Results were summarized in an analysis of the strengths, weakness, opportunities and threats (SWOT) to show advantages and challenges the company faces when introducing the circular economy concept. To better understand the operational processes, while visiting the recycling factory, a variation of a multi-moment recording procedure was conducted assessing the procedures of a particular operation occurring at a specific place over a specified duration to capture how the work was completed and in what ways (REFA-Methodenlehre 1978). The goal of this procedure was to compile a detailed description of the actual processing steps of all the work in the recycling facility, to discover challenges and risks. This chapter describes the effect of developing and implementing a circular economy approach at EF. The results indicate that a circu-

S. Weber (✉)
School of Environment, Resources and Sustainability (SERS), University of Waterloo, 200 University Ave W, Waterloo, ON N2L 3G1, Canada
e-mail: sabine.weber967@gmail.com

S. Weber
School of Fashion, Seneca College, 1750 Finch Ave E, Toronto, ON M2J 2X5, Canada

Present Address
S. Weber
446 Drake Circle, Waterloo, ON N2T 1L1, Canada

© Springer Nature Singapore Pte Ltd. 2019
M. A. Gardetti and S. S. Muthu (eds.), *Sustainable Luxury*,
Environmental Footprints and Eco-design of Products and Processes,
https://doi.org/10.1007/978-981-13-0623-5_7

lar economy approach in the luxury fashion industry is possible and is beneficial to extended business, can generate revenue and reduce environmental degradation. This chapter explains that a circular economy approach in luxury fashion requires a company to produce luxury products that are timeless, high quality, durable and retail at a high price.

Keywords Circular economy · Luxury fashion · Entrepreneurship · Textile recycling · Take-back · Reuse · Innovation · New business models

1 Introduction

Today's luxury fashion is presented on the catwalks and in the glossy pages of glamour magazines as dream-like visions. However, there is a real effect of fashion on the environment, and luxury fashion companies are increasingly cognizant of the substantial need for sustainable innovation.

One of the luxury fashion brands that has taken on a leading role in environmental justice and sustainable development is Eileen Fisher, New York (hereafter EF). The company is committed to minimizing waste and developing a sustainable, low-carbon, resource-efficient and competitive business. Human rights and sustainability are part of everyday decisions and not limited to special projects or specific products.

The company launched its Green Eileen initiative in 2009, now called Eileen Fisher Renew, to offer a take-back programme for used EF clothing. This unique programme sorts gently-used items and cleans them professionally before reselling them in Renew stores. The programme is sustainable because it extends the lifespan of the textile product through further use and represents an entrepreneurial interpretation of extended producer responsibility since the company sells its products twice. If garments come back damaged and are not good enough to be resold under the Eileen Fisher Renew programme—because they are stained, have holes or show strong signs of use—the company prepares them for reuse with various techniques including mending, overdye,[1] resewing and felting. This work is done in its new factory in Irvington, New York, focused on reuse and remanufacturing. In 2016, the factory processed about 170,000 pieces, accounting for about 2% of the total annual EF production.

Since there are few luxury fashion companies which follow a circular economy model, a case study approach was chosen to see and describe how the company has sought to achieve a circular economy model in the luxury fashion industry. EF represents a financially successful fashion business model committed to its values, these are: environmental sustainability, human rights, women and girls empowerment

---

[1]Eileen Fisher uses the term *overdye* to mean the process of redying garments that have already been dyed (redying means dying a second time). This process is typically done to conceal stains and other discolourations from extended use.

(Director of Social Consciousness)—the company became a certified B-Corporation[2] in December 2015 (Fisher 2017). Therefore, the company offers a useful case study to observe the opportunities and challenges of a circular economy model. Since the examination of a circular economy requires a holistic approach, one that considers the entire life cycle of each garment, this chapter explores multiple issues related to a circular economy, such as reduce, reuse and remanufacture or recycling.

2 Map of Chapter

This chapter will first present a literature research on the current fashion system, as well as circular economy approaches to the fashion industry that integrate reduce, reuse and recycling (see Sects. 3–6). Next, it will present the results of semi-structured interviews with employees from EF, representing different departments and operating at various functions in the company (see Sects. 7 and 8). It will introduce the Eileen Fisher company (see Sects. 9 and 10). Next it will show the findings of an observational study of EF's recycling factory, with special attention paid to its operational practices (see Sects. 11, 11.1, and 12). The objective of the chapter is to reflect multiple perspectives of a circular economy approach and to collect information about this company's progression towards a circular economy. Attention will also be paid to achievements and challenges that remain (see Sect. 13). In the context of the luxury fashion industry, there is a lack of research on the key features of a circular economy model. The goals of the chapter are to show how a circular economy model can be put into practice and to establish a guiding conceptual framework for other luxury fashion brands seeking to transform business towards a more circular economy.

3 The Current Fashion System

The current fashion system in North America is a linear economy predicated on buy, use and disposal, with little diversion or reuse of unwanted textiles. Modern consumers purchase more garments than any other period. According to the American Apparel and Footwear Association, the average consumer purchases 62 garments per year (The American Apparel and Footwear Association 2012). Garments are frequently worn during the first year when purchased, but already in the second year, fewer than half of garments are still worn on a regular basis (Smith 2012). A study in the Netherlands found that the average consumer gets rid of its garments after three years and five months regardless of use (Uitdenbogerd et al. 1998). While the

---

[2]B-Corps are for-profit companies that are determined to conduct business as a source for good. The non-profit B Lab is certifying these businesses in regard to social and environmental performance, accountability and transparency (B-Corporations).

use phase for garments is short, a garment is worn an average of seven times before disposal (Maybelle 2015). The result of this consumption and disposal is an average of 37.2 kg of textile waste per person per year (Council for Textile Recycling 2014). Textiles make up between five to ten per cent in landfills in North America (Jensen 2012; United States Environmental Protection Agency 2013) producing carbon dioxide (CO_2) and methane emissions and contributing to climate change (Fletcher 2013).

In North America, about 85% of all post-consumer textile waste ends up in landfills (Council for Textile Recycling 2014). These include unwanted garments and home textiles, such as towels, beddings, and drapery plus shoes and accessories. Only 15% of all textiles are diverted from the waste stream through textile donations, and only three per cent are recycled and reclaimed into new fibres. Four and a half per cent are recycled and converted into new products such as desk counters, insulation or stuffing materials (Council for Textile Recycling 2014). These numbers show that the current fashion industry is based on a linear business model of buy, use and dispose.

Why must the fashion system transition towards a circular economy? The current business model of the fashion industry relies on limitless growth, but the Global Fashion Agenda describes in its Policy brief, "A Call to Action for a Circular Fashion System", that the current fashion business model has reached its physical limits (Watson 2017). Continuous population growth will increase the global fibre consumption, leading to a global fibre gap and a continuous increase of the environmental burden. As the Global Fashion Agenda puts it, "With the world's population expected to exceed 8.5 billion people and global garment production to increase by 63% by 2030, this [economic] model is reaching its physical limits" (Watson 2017, p. 1). To solve the compound issues of scarcity and environmental degradation requires a shift towards a new sustainable development model, towards a circular economy. Therefore, a circular economy is seen as a "key driver for sustainability" (Bocken et al. 2016).

4   Challenges for a Circular Economy in the Fashion Industry

While definitions of the term "circular economy" vary, this chapter follows the concept offered by the industrial ecology or eco-industrial development. This approach seeks to decouple economic growth from environmental depredation by suggesting that economic development can be achieved in harmony with the environment by reducing waste and increasing resource efficiency. Geng and Doberstein describe a circular economy as "realization of closed-loop material flow in the whole economic system" (Geng and Doberstein 2008, p. 232), which Lieder and Rashid (2016) describe as "a solution for harmonizing ambitions for economic growth and environmental protection" (p. 37). In other words, a closed-loop material flow of garment production, in which garments are endlessly reused and recycled into new products, decouples economic growth from environmental pressure. In such a system, waste is seen as a by-product of manufacturing processes and used as a resource for other industries (Geng and Doberstein 2008). Likewise, products at their end of life cycle are remanufactured or recycled to become the feedstock for new products. A circular economy aims to design "an industrial economy that is restorative or regenerative by intention and design" (MacArthur 2013, p. 15). Ghisellini et al. explain that a circular economy includes "the design of radically alternative solutions, over the entire life cycle of any process as well as at the interaction between process and the environment and the economy in which it is embedded" (Ghisellini et al. 2016, p. 12). A sustainable economy will require a greater focus on solving textile waste generation and appreciating resource scarcity. While this approach might lead to the assumption that a circular economy is a waste management approach which mainly treats the symptom of waste, a circular economy approach also includes waste prevention. Indeed, the majority of literature on the topic of a circular economy is based on the waste management approach of the 3R's: reduce, reuse and recycle (Ghisellini et al. 2016). Frequent use and long product life would optimize the value of products and could lead to a reduction of consumption. In an ideal system, materials would circulate infinitely. However, this requires a radical shift in the business operations of the fashion industry.

The common business model in the fashion industry is driven by consumption not by reduction, while the constant desire to change clothing has opened new business opportunities for fashion brands by forgoing the desire for sustained ownership and by making the idea of "temporary owning" popular. Instead of purchasing garments, consumers are renting or borrowing clothes for a fee. Uche Okonkwo (2016) explains that the concept of "lending" clothes has been practised in luxury fashion for decades: "when brands were "lending" clothes and accessories to celebrities for special red carpet events like the American Oscar Awards" (Okonkwo 2016, p. 232). Temporary ownership represents one form of performance economy, which Stahel explains "goes a step further [than a circular economy] by selling goods (or molecules) as service through rent, lease and share business models". Although a promising concept, the performance economy is beyond the scope of this chapter.

5   Cultivating a Circular Economy in the Luxury Fashion Industry by Reducing Consumption

There are multiple ways to cultivate a circular economy in the fashion industry. One that offers promise would be to reinvest in luxury fashion since the luxury fashion industry has traditionally aimed to produce clothing intended for frequent use and long product life, with garments that are durable and timeless. Phau and Prendergast outline four attributes of luxury products: brand identity, quality, exclusivity and customer awareness (Phau and Prendergast 2000, pp. 349, 361). Instead of selling a high quantity of mass fashion, the luxury industry aims to sell fewer garments with higher retail prices, therefore, reducing consumption and increasing sustainability. However, the business practices of contemporary luxury fashion are increasingly coming more in line with fast-fashion than sustainable fashion. Sull and Turconi describe fast-fashion as a "retail strategy of adapting merchandise assortments to current and emerging trends as quickly and effectively as possible" (Sull and Turconi 2008, p. 5). Most successful fashion retailers and brands operate under a fast-fashion strategy, relying on fast-changing fashion cycles and trends and not because they are selling timeless clothing. To keep up with this fast-changing fashion cycle, some luxury fashion brands have even adopted a fast-fashion production cycle. For example, in 2016 Burberry made headlines when it announced that it will show their collections twice a year but that collections can be only bought directly after the show based on the concept: "buy-now-wear now" (Conti 2016, p. iv). The result of this rapid speed to the market is that consumers can immediately purchase the product because it is already produced, rather than waiting the customary six months for garment delivery (Conti 2016). According to Okonkwo (2016), the economic, social and technological aspects of the fast-fashion industry have led to profound transformations in the luxury fashion industry. While "mass fashion brands have attuned their business strategies to resemble those of luxury brands and now offer similar goods at a lower price", changes in investment structures in the luxury sector have "increased pressure on luxury brands, for rapid sales and profitability" (226). These changes raise the question whether luxury fashion will reduce consumption.

The features that characterize fast-fashion are new merchandise each week, highly responsive and flexible supply chains, rapid speed-to-market, highly sensitive to catwalk trends and the perception of product scarcity (Sull and Turconi 2008). For many years, fast-fashion was only associated with mass-retail chains like H&M, Zara or Forever 21, but luxury fashion labels are strongly influenced by fast-fashion given consumer desire to follow the fashion trends. As a result, "luxury consumers now follow fashion trends religiously [based on] their desire to be trendy at all times and in some cases, at all costs" (Okonkwo 2016, p. 233).

While fast-fashion is highly fashionable clothing, it requires constant replacement due to new fashion trends and shifts in the fashion identity of consumers (Smith 2012). Fast-fashion clothes which are worn only a few times (Birtwistle and Moore 2007) and have led to a consumer habit of "disposable" or "throwaway" fashion (Bhardwaj and Fairhurst 2010). Luxury fashion differs in several important respects. Luxury products stand for high-quality and timeless products, but luxury also represents innovative, creative product design and always retails at a premium price. However, under investor pressure to adopt fast-fashion production cycles to maximize profitability, luxury fashion brands have introduced mass marketing retail strategies into the luxury fashion scene, in the process creating three distinct categories of luxury fashion: "fashionable luxury", "common luxury" and "true luxury" (Okonkwo 2016). Fashionable luxury refers to the fast-fashion method of luxury fashion dependent on ever-shortening cycles, while common luxury refers to cheaper, trendier and more visible styles than true luxury, which refers to timeless, durable and rare fashion. As a concept, luxury fashion does not necessarily reduce consumption. Nevertheless, it is more sustainable than fast-fashion given how garments are kept and managed at the end of their life. While typical fast-fashion garments will only be kept for a short time before being thrown away, luxury fashion, even (fast) fashionable luxury garments will be kept because of the perceived and material value of the garment (determined by factors like brand recognition and garment price).

5.1 Influence of Price in a Circular Economy Luxury Fashion System

The price of a garment plays a determinative role in the duration and longevity of its use. In general, consumers keep garments longer when the investment value was high (Bye and McKinney 2007). As Okonkwo puts it, "Consumers literally cannot afford to adopt the "throwaway fashion" attitude towards luxury goods" (Okonkwo 2016, p. 231). Further, used luxury products remain valuable due to the intrinsic value of the brand. Indeed, even in their afterlife, consumers sell their luxury clothes for substantial amounts of money (Okonkwo 2016). Additionally, consumers might also donate or pass on their clothes to friends and family. A study from British Columbia, Canada, found out that the main reason participants gave for donating their materials was to help others (37%) to reduce clutter (22%) and to find a good home for valued goods (16%) (Vancity 2016, p. 1). This consumer desire to own, keep and responsibly manage valuable products offers opportunities for sustainable business practices.

6 Two Common Circular Business Models

Stahel distinguishes two circular economy business models: The first promotes reuse and the extension of a product's life cycles while the other focuses on recycling unwanted materials into new products. In both, the key to success is the people's willingness to transform the system (Stahel 2016) and the common objective is "to maximize value at each point in a product's life" (Stahel 2016, p. 436). The value of products can be optimized through frequent use and long product life. Products must be durable, timeless and repairable to increase opportunities for reuse, and must also be easy to remanufacture and recycle.

6.1 Reuse–Rewear

A 2015 study conducted in Ontario, Canada, found that consumers seek to prolong the value of their products. The most common practice to get rid of unwanted garments is to donate them. Most participants, 92%, know at least one place of a donation box or are aware of the opportunity to arrange a pick-up. Roughly 18% of the participants swap their used clothing, 38% have tried at least once to resell a garment, and a further 12% have used take-back programmes. In fact, take-back programmes are relatively unknown in Canada. Only 31% of participants have admitted to knowing of a retailer offering a take-back programme, but 70% of participants think, take-back programmes are a great idea (Weber 2015, pp. 72–73). This result suggests that customers are interested in possibilities for extending the value of their clothes in ways that would benefit the wider implementation of take-back programmes. Although participants had claimed that the time and effort required to resell a garment kept them from practising it more often, second-hand shopping for clothing has become a fashion trend. A study from British Columbia (BC) confirms the growth of the second-hand economy in BC, with clothing and shoes the most popular items of the second-hand business (Vancity 2016).

6.2 Recycling

Although all textiles can be recycled in one or the other way (Stall-Meadows and Goudeau 2012), worn textiles are not commonly used as a source of raw material for new products, i.e. recycled into new fibres—in North America or even worldwide. According to the Council for Textile Recycling, only 3% of the collected and diverted Textiles in North America are being recycled into new fibres, and 4.5% are being recycled and converted into other products (Council for Textile Recycling 2014). There are some opportunities for ragging, usually only for garments made of natural fibres, and shredding, but those mechanical recycling opportunities mainly produce products of lower value such as insulation and stuffing materials and are therefore a down-cycling. This kind of recycling can also be called an open-loop recycling process, because the new products aren't remade into yarn or fabrics and are therefore not "closing the loop". Only in a closed-loop system will the material flow continue infinitely because the recycled material becomes the source of a new fibre (this is also referred to as fibre-to-fibre recycling). If natural fibres are recycled in a mechanical process, the fibres become shorter, which requires a mixture with virgin products to produce a decent yarn quality (Fletcher 2013). Research into the textile recycling processes for materials at the atomic level has increased in the last few years, seeking to achieve closed-loop recycling. For example, the company Evrnu has developed a technology to dissolve natural fibres into cellulosic pulp which can be respun into high-quality fibres, but the technology is not yet available on a larger scale.

Today, chemical processes can recycle synthetic fibres without any loss of fibre quality and producers of recycled polyester and nylon claim that these materials require up to 30% less energy in their production (Koh 2017; Victor Group Inc. 2008). Nevertheless, the scale is small and producers are not required to determine and explain the source of recycled material. Consumers can purchase recycled polyester, but it is unclear what percentage is recycled polyester or whether the recycled polyester is made from post-consumer textile waste or recycled plastic bottles. However, with an increase use of recycled fibres in the fashion industry, standards are becoming more important. Organizations like the Textile Exchange and the Scientific Certification Systems (SCS) provide independent third-party verification on the input materials if producers wish to have this information provided (SCS Global Services 2014; Textile Exchange 2017). While the SCS Certified Responsible Source™ standard evaluates products made from pre-consumer or post-consumer material diverted from the waste stream, the audit provided by the Textile Exchange for the Recycled Claim Standard and the Global Recycled Standard requires additional verification of the source of recycled materials at the recycling stage (SCS Global Services 2014; Textile Exchange 2017). Nevertheless, the biggest challenges in textile recycling remain to scale the existing recycling methods and to develop recycling processes for fibre blends (Koh 2017) and other types of material, such as spandex materials, polyvinyl chlorate (PVC) and polyurethane (PU). Further challenges for all chemical textile recycling processes are the unknown chemicals added to each garment (American Chemical Society 2017).

7 Significance of This Study

Literature on sustainability and fashion has explored opportunities and barriers against sustainability in the fashion industry and its supply chain (Black 2012; Fletcher 2013; Henninger et al. 2017; Karaosman 2016; Moon et al. 2015; Pedersen and Andersen 2015; Pedersen and Gwozdz 2014; Strähle et al. 2015). Few studies have focused on textile waste from a consumers perspective and their disposal behaviour for clothing (Birtwistle and Moore 2007; Laitala 2014; Lang et al. 2013), and some studies explored textile recycling possibilities (Hawley 2009; Zamani et al. 2015). Literature on a circular economy approach is often based on design possibilities in the fashion industry (McCourt and Perkins 2015; Smith et al. 2017), and there is less research on post-retail responsibilities of brand owners to take garments back (Kant Hvass 2014), but there is no research on such an approach in the luxury fashion industry. According to Hvass, post-retail responsibilities are an emerging field in the fashion industry with limited best practice studies (Kant Hvass 2014) and only the Ellen MacArthur Foundation's report, "A New Textiles Economy: Redesigning fashion's future" released in November 2017 outlines a vision for a circular fashion system but claims system-level change, collaboration, and innovation are necessary to make the fashion industry more sustainable (Ellen Mccarthur Foundation 2017).

8 Methods

The company EF offers a unique case to observe the opportunities and challenges of a circular economy model. Since the examination of a circular economy requires a holistic approach, multiple issues are explored related to a circular economy, such as how a company seeks to integrate reduce, reuse and remanufacture or recycling.

In the context of the luxury fashion industry, there is a lack of research on the key features required to scale a circular economy.

A case study method was adapted from Curwen et al. (2013), who used a qualitative method involving in-depth interviews and direct observation adopted from Yin (2013) to conduct a sustainability study on EF's apparel product development.

In November 2017, twelve semi-structured interviews were conducted with employees from EF, representing different departments (see Appendix 2) who operate various functions in the company. Additionally, respondents were contacted via email and telephone to review answers and provide further clarification. Their responses were analysed to outline EF's approaches to both luxury fashion and a circular economy. In-depth interviews generated rich descriptions that were analysed according to a content analysis method. Further, two EF stores and the recycling factory were visited, and additional data from the company was obtained.

For some of the participants in the study, questions were prepared ahead of time (see Appendix), while for others, questions naturally emerged during the process of the facility visit and the subsequent multi-moment assessment. Questions for the site visits were motivated by an aim to understand system processes related to working steps necessary to sort, wash, deconstruct, store, remake, mend and/or felt. In short, this study sought to understand how and why EF sought to implement a take-back programme.

For the in-depth interviews, the types and topics of questions asked depended on the role and knowledge of the respondents. A qualitative codebook was generated from a review of the literature to guide analysis. Content analysis was driven by three fundamental concerns. The first was the challenges and strengths for Eileen Fisher to become circular. Secondly, how "closing the loop" affects various departments—particularly design and retail. Thirdly, how the take-back programme was developed and how the company innovated. This codebook was used to categorize responses into distinct topics that allowed for comparison between responses.

This study explores how EF has developed its take-back programme and how this programme led to the development of recycling operations at EF. Results were summarized in an analysis of the strengths, weakness, opportunities and threats (SWOT) to show advantages and challenges the company faces when introducing the circular economy concept.

While visiting the recycling factory, an evaluation of its processes was conducted. The mode of inquiry was a variation of a multi-moment recording procedure, a method which assesses procedures of a particular operation occurring at a specific place over a specified duration that captures what work was completed and in what ways (REFA-Methodenlehre 1978). The goal of this procedure was to compile a detailed description of the actual processing steps to discover challenges.

Interview participants listed by job titles were

- Director of Social Consciousness
- Manager of the Eileen Fisher reuse programme
- Wholesale Marketing Manager
- Buyer for EF retail stores

- Designer for knitwear EF collection
- Designer for Resewn collection
- Facilitating Manager for Fabric R&D
- Sustainability Leader
- Recycle operation manager
- Three sales associates

  Sites visited

- EF headquarters in New York, NY and Irvington, NY
- "The Lab" in Irvington, NY
- EF store in New York, NY
- EF recycling factory in Irvington, NY

9 The Eileen Fisher, Inc. Enterprise

Since its founding in 1984, Eileen Fisher, Inc. has become a $500 USD million enterprise which employs 1200 people. Based in Irvington, New York, Eileen Fisher sells women's business and casual clothing as well as accessories and shoes in its stores throughout North America. Eileen Fisher garments are also offered in specialty stores and department stores in North America, while expanding its international presence to England, Dubai, Kuwait, Kazakhstan, Turkey and Thailand. New garments typically retail $100 USD for tanks and camis; $150 USD for tops and tees; $350 USD for sweaters and cardigans; $400 USD for jackets and vests; $200 USD for pants; $180 USD for skirts.

10 Profile of the Company and Its Founder

Eileen Fisher started her career working as an interior designer in New York, where she discovered Japanese aesthetics. She was fascinated by the simplicity and the timeless designs of kimonos. Feeling overwhelmed by fashion trends in the eighties, with the constant change in colours, fabrics and shapes, she started designing clothes which she would like to wear. The result was a line of clothing based on simple shapes, comfortable fit, charming to the touch, with timeless colours. The garments were designed to exist outside of fashion trend cycles and were easy to mix.

Her strong vision drives the company towards achieving social and environmental justice. As a company, EF's sustainable approach extends to its business operations and corporate culture. Already in the early 1990s, the company recognized the need to commit and improve human rights conditions among its global supply chain partners, and factories had to introduce international workplace standards for their facilities. A few years later, the company extended this commitment towards sustainability with a clear obligation towards the environment. The company is motivated by a relentless

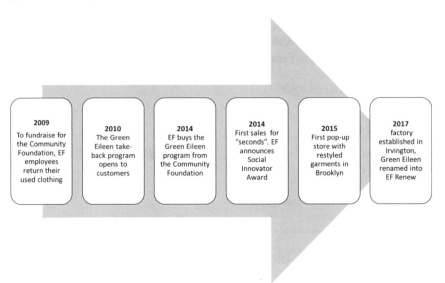

Fig. 1 Milestones of Eileen Fisher take-back programme (figure created by author)

push towards innovative and sustainable methodologies of design and reuse. In 2005, Eileen Fisher sold parts of the company under a staff ownership plan to her employees. Today, 40.5% of the company is owned by its employees, with this step Eileen Fisher wanted to ensure that the company would remain in the hands of people and not be sold to investors only interested in the bottom line. From its beginning, the company supported the movement of B-corporations and received B-Corp certification in 2015. The declaration on independent of B-Corps says: "[Conducting business] is purpose-driven and creates benefits for all stakeholders, not just shareholders" (B-Corporations).

EF has pioneered sustainable, true luxury fashion in the industry, and offers a productive model for how the industry might move towards a circular economy. Perhaps the most notable aspect of EF as a luxury fashion company is its enthusiasm and determination to take-back, remake, recycle and resell their products (Fig. 1).

10.1 The Birth of EF's Take-Back Programme—Collecting Garments for Reuse

Around 2008/2009, EF founded two unrelated and separate foundations: one is a private family foundation, the other one a community foundation. By definition, a community foundation has to be funded in part by the community or by other kinds of funders such as other foundations or businesses. Over the next two years, Eileen Fisher and her team held several meanings to discuss sources of income for the community foundation. One employee later recalled the tenor of the discussion as creative and exploratory: "What if we did this? What if we tried this?" An essential question was asked: "Why not collect our beautiful clothing that we have had now for a long time and sell it? Maybe that income can provide money for the foundation". Since the company has a very generous clothing allowance, employees had accumulated a lot of clothes in their closets. The programme started with just employees bringing their own, still in perfect condition and dry-cleaned, clothes, and selling them in their store in Irvington. The store manager was a big supporter of this initiative and the used clothing sold well.

Later, some of the customers expressed their interest in wanting to participate in this initiative. The funding initiative became the Green Eileen Programme. After they began to accept clothing donations from customers, the programme quickly grew. "When we opened it up to customers we were totally flooded with clothes," one employee recalls, "Mountains and mountains of clothes". There was soon more donated clothing than could be processed. This volume required additional storage and retail space. Although the initiative began with a rack of used clothing in the store, soon a little section of the store was apportioned for used clothing. Eventually, the programme grew so successful that two standalone stores were added. At this time all profits were going into the Community Foundation.

The accumulation of thousands of used garments led to a crisis for EF. The company donated some of the garments in good condition, or gave them to local artists working with fabrics, but most of the returned material was stocked. The surplus of material, stored for years, incurred additional costs. The manager of the reuse programme remembers: "We didn't have any big sellable solutions for it. We thought it was responsible to hold on to it". Such challenging goals can be termed as organization-wide enabling condition for innovation (Nonaka 1994). Its certified status as a B-Corporation meant that EF was not obligated to maximize profit, which provided the potential to store those garments while determining a more viable long-term solution. EF also relied on its values rooted in sustainability and innovation to guide its solution.

10.2 The Beginning of EF's Recycling Operations

In 2014, EF bought the Green Eileen programme from its foundation. With the increase of the programme, new challenges arose. Thousands of garments were donated, but not all of them could be resold in perfect quality. Many garments were not in perfect condition, but were still good enough for reuse, so-called Seconds. There were also many garments which required repair and recycling. With thousands of garments accumulating in storage, the employees of the Green Eileen team started thinking about recycling opportunities for the material.

Firstly, additional sales were established for "seconds". During these sales, slightly damaged used garments were sold at low-price brackets: $15, $20 or $25. These sales attracted so many new customers, including some perhaps who would have otherwise never considered an EF garment within their budget, that these sales have been maintained as biannual events in Seattle. One employee reported: "During those sales lines of women are waiting for hours to buy a piece". According to information from EF during the last sale, it sold about $30,000 in four hours, roughly 1,500 pieces in four hours, or one garment sold every 0.16 s. Such results can only be achieved with luxury products, which maintain a high degree of desirability, even in a "heavily used" condition. The desirability of luxury fashion, dependent on their ability for reuse, makes them the ideal material for a circular economy.

The second outcome was the Eileen Fisher Social Innovator Award. In 2014, EF partnered with The Council of Fashion Designers of America, Inc. (CFDA) a not-for-profit trade association, to implement a one-year long training and mentorship programme for three post-graduate fashion designers. During the programme, graduates are placed on rotation at EF and work collaboratively on sustainable design challenges. In 2014, the contestants created a 500-piece collection out of returned EF clothes. One employee stated: "We challenged them to make it sellable, to make it beautiful, to make it profitable." The collection was finally sold during a pop-up store event in Brooklyn. The project created tremendous excitement among staff and customers, and all involved in the project were eager to continue its development. Three techniques were established as possibilities to extend the use of the textile material and to scale the programme: cut it open and sew it back together, overdye it with natural colours, or felt it. Those techniques are the base for the EF Remade collection. While the EF team had plans to start a factory even before the CFDA Social Innovator programme, the programme shifted EF's emphasis towards the Resewn line. Eventually, EF decided to focus all the factory's efforts on the Resewn line rather than on mending and repair. Nevertheless, simple garment defects continue to be mended and repaired for further reuse.

Rather than adhering to a formal business plan, developments in this context were incremental, adaptive and responsive to emergent problems in the system. Solutions were collective, innovative and reactive.

10.3   The Business Case

In 2017, a factory was established next to EF headquarters in Irvington, New York, to further explore and develop repair, resew and felting techniques. Emphasis was also placed on finding ways to make factory operations financially viable and scalable. EF also established a recycling centre next to the retail store in Seattle, which focuses on the overdye technique. Around the same time, the "Green Eileen" programme was renamed into "Eileen Fisher Renew" and the EF Lab Store became the first EF store showing the full circular economy concept of the company to its customers.

In 2016, the company received about 170,000 pieces back, about 2% of its total annual production (see Fig. 2), which means on average about 3000 garments per week arrive at the two recycling centres in Irvington and Seattle. About half of all garments taken back are in such perfect condition that they can be cleaned and resold;

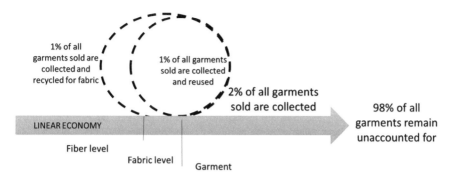

Fig. 2 Scope of Eileen Fisher take-back programme (figure created by author)

the other half requires recycling. To process all these garments and to coordinate all activities around the circular economy, a new department was established. The "Renew team" consists of about forty employees across two recycling centres, two dedicated retail stores and the Factory, including a designer from the Social Innovator Award programme.

11 What Factors Contribute to or Challenge EF's Circular Economy Approach

The interviews were analysed to determine the strengths, weakness, opportunities and threats (SWOT) of EF's circular economy approach. Five key aspects were identified as central to becoming successfully circular: the concept of a true luxury product, garment design, garment manufacturing, distribution channels and customer base. The following section discusses how these aspects have benefited or hindered EF's approach, as well as how the company has sought to address and overcome internal challenges and external threats.

1. The concept of a true luxury product: EF clothing is not about trend, but about style, garments are timeless and can still be worn years after purchase—the essence of true luxury. The products reflect the company's environmental and social responsibility values. EF garments are made of highest fabric quality, by artisans whose technical craft makes it possible to wear the product for years. The result is beautiful, functional, durable clothes. Products are optimized for frequent use and long product life. The buyer explains that the "[EF] customer typically likes the more timeless simple pieces that have a little bit of an edge". The retail price is high, so consumers who can't afford a new EF garment are still keen on a second-hand piece since the brand seen as desirable. Since garments are made of high-quality material and are durable, it is possible to use them for a very long time, and this is particularly important when the garment should be sold a

second time. The luxury even extends to the reuse line: take-back manager claims that the garment resold are "in perfect condition. We have very high standards."

2. Garment design: EF garments are unique fashion products in that they have minimal accessories and are loose-fit—unique from an aesthetic standpoint, and vital for their remaking and recycling. The minimal use of accessories also decreases the risk of product damage (lost buttons, broken zippers), which increases product longevity. EF garments are comprised of wide garment panels with higher than usual yardage compared to narrow-fit clothing, so that fabric pieces are easier to remake into new garments. According to the knitwear designer of EF's main fashion line, "the more seams we put into something, the less yardage they can use if they're going to cut and resew something". The knitwear designer is conscious and sensitive to the use of seams and accessories, and considers how her garments can be more easily remade.

The desire for durability and longevity is encoded into every aspect of design, starting at the sourcing of material.

Design at EF is not limited to the garment, but includes the "material development" which encompasses fabrics and yarns. For example, it includes weave and pattern development with mills, print and other surface design as well as yarn and stitch development. The company's vision is to use only sustainable fibres by 2020 (Fisher 2017). This means the garment designers seek to use reclaimed fibres for their collection, for example reclaimed cashmere fibre, as well as recycled cotton and wool, or recycled polyester and nylon while the material development team works with suppliers to achieve high-quality materials with reclaimed fibres, natural fibres especially, until they can withstand the high requirements for an EF garment (Manager for Fabric R&D). However, there is a lack of available technology to recycle fibre blends when the material is broken down to the monomers, limiting EF's capacity to source recycled fibre material. To overcome this external challenge threatening all fashion companies, EF has joined a brand consortium called the "Circular Innovation Working Group" which has been organized by Cradle to Cradle Fashion Positive Institute. Therefore, EF is funding due diligence research into the viability and potential to develop and scale up synthetics (Manager for Fabric R&D). The Director for social consciousness further explains: "We are looking outside the company for a solution to these very complex fibre challenges that we have". EF remains optimistic about future applications for sustainable design: "Each one of [the external companies] is entrepreneurial, so we are, in effect, funding external entrepreneurs to work on their own fibre technology; perhaps one of them will be the solution we're looking for, we don't know yet."

3. Manufacturing new clothing: Having full control over the supply chain can be a strength for a company because it allows for the highest level of transparency and offers the most possibilities to improve. EF outsources most of its production capacity. The director of social consciousness explained that it is difficult for EF to be certain about every production practice, although all manufacturers are nonetheless long-time partners. This sustained relationship provided the opportunity to collaborate towards achieving EF's values. In one instance, a production

had produced huge amounts of offcuts (leftover material that results when garments are cut in piles) in a natural colour. To keep this material from the waste stream, all fabric scraps were opened into yarn and knitted into new fabric. While this was a successful project, the company is not involved any further in how their partners manage fabric scraps or other waste. While leftover yarns and fabrics were sold at the Lab Store, this operation is very small and cannot absorb all the leftover materials caused by the company and its partners.

Manufacturing used clothing: In addition to the lack of fibre recycling technologies available for used textiles, there is little remanufacturing knowledge for used textiles. Partly because the trend is so new, and the field is still emerging, the lack of skilled remanufacturing labourers and sites has led EF to open their own recycling facility which aims to develop and train remanufacturing knowledge and skills, while this is a huge step to become a circular economy, it is necessary to develop best practices to remanufacture garments. Interestingly, according to the take-back programme manager, the Resewn garments "don't look remade. They just look like brand new clothes. So, in my mind, this could be something where we actually sell it in our stores, and you don't even make a deal out of it being remade. You just sell it". However, although the products made under the Resewn programme are of high value, the remanufacturing and the repairing isn't profitable yet, in contrast to the reuse programme which is profitable.

4. Distribution channels: EF owns 68 retail stores globally (including two Renew stores, four stores in the UK and three stores in Canada), this provides EF with enough retail space to showcase their circular economy model, to take used garments back (only in its own retail locations within the USA) and to sell reused garments. The director of social consciousness describes the opportunities of business expansion through the circular economy model through different price points for reused or restyled clothing: "It's ... about growing business reaching out to new customers." One of the ways EF has sought to do so is by cultivating a vintage style as a means of interesting younger and less affluent customers to purchase reused EF clothing. As the director of social consciousness puts it: "our price point is high, so for significantly younger people probably it's not as likely for them to buy into it, but they would buy re-worn, remade clothing because it's popular right now. The vintage is popular". The manager of the EF reuse programme explains that while many of EF's wholesale partners are interested in showcasing their Renew line, many operational hurdles remain. Most retail partners are not part of the take-back programme, do not sell EF's reuse collection, nor do they emphasize or prioritize EF's circular economy model. To overcome this barrier, EF has implemented special sales events with their partners to communicate their circular business model and products. During one of these events which occurred at five Nordstrom locations and online in September 2017, Nordstrom sold the main line of EF clothes along with the renewed and the remade EF clothing lines. As the wholesale marketing manager explains, "This is the first time we were able to tell the [sustainability] story in a wholesale environment".

5. Customer base: The success of the take-back programme depends on the customer. EF has cultivated a relationship with its clientele who share and support the

company's values. Their respect for the brand extends to respect for its clothes. In other words, the people who buy EF clothes are also the people who take care of EF clothes and who treat their garments well. These factors combine to facilitate a successful take-back programme. More than 50% the received garments are in perfect condition according to the manager of the EF reuse programme. Part of this relationship can be attributed to the design ethos of the company. The EF knitwear designer explains that the luxury aspect of the brand derives from "the thoughtfulness that goes into every part of the garment, and the time we take to make sure the thing is curated in a way that's thoughtful. So anybody that the fibre passes hands through, the factories, the workers, we want to make sure that we're putting out a product that is the best we can do, and the nicest that we can do it". While EF has a strong and loyal customer base, it is also limited to women who can afford these garments. Selling the EF remade and especially the renewed collection has allowed the company to extend their business and reach a new customer group. The wholesale marketing manager explains: "we did see a younger customer reacting to the story. Whether or not they bought the product is still here and there, but they definitely felt like the story was something that resonated with them". The marketing manager further explained that the promotion with Nordstrom in September 2017, attracted approximately 2000 new customers in a two-week time frame. Further, developing a luxury company towards a circular economy extended the brand recognition and its customer base.

A SWOT analysis of EF investigates how the company has addressed the circular economy approach in luxury fashion apparel. Since the company values true luxury as a business concept, quality, timelessness and durability guide the design and manufacturing of its products. The relationship between the company and its customers plays a significant role in this business model. EF's business concept relies on the customers sharing the same values as the company. The take-back programme depends on the customer to bring garments back, but also to purchase used garments or garments made of reclaimed fibres. Loyalty and trust provide the basis for this successful relationship. Challenges include global supply chains, outsourced manufacturing and variable external sales partners, but EF has sought to address any potential issues through transparency and special projects. In the long run, however, supply chain and sales partners must be fully integrated into the circular economy business model. The main threats to becoming circular are the lack of knowledge in the textile and apparel production sector. The fibre recycling and textile remanufacturing industry remains a niche. EF has recognized this need and has taken on ownership of the textile remanufacturing challenge by developing its own facility while collaborating with others to invest in research for textile fibre recycling. While the EF programme reflects a financially successful and highly innovative business model overall, there are nonetheless challenges in the daily operations in the recycling factory that suggest further opportunities for innovation.

While the SWOT analysis provides insights about the key factors which need to be considered when implementing a circular economy approach, the next section explores the company culture at EF in greater detail and considers the factors that enable the company to create and explore sustainable practices.

11.1 Entrepreneurship and Innovation

Under the North American Industry Classification System (NAICS), EF is listed as a manufacturer, but over the years the company outsourced most of its production to China, India or Japan, depending on the required production techniques such as batik, Shibori or hand knitting. For example, in Peru the company follows a fair trade production model, crafting luxury fully fashioned knitwear sweaters (Curwen et al. 2013). EF has invented a manufacturing method for processing used clothing, an exceptional development in the North American fashion industry. In fact, few facilities in North America are capable of remanufacturing used clothing. Few external experts exist with a background in textile remanufacturing, and no machinery exists specifically suited for the remanufacturing processes. As a result, the whole operation is working on a "learning by doing process", developing its own best practices: generating knowledge and developing its own machinery based on this knowledge. For example, the company has developed the first felting machine for used clothing based on its knowledge of trying to felt fabric scraps and garment pieces. In so doing, EF has become an entrepreneur in textile remanufacturing.

Ikujiio Nonaka describes innovation as follows: "A process in which the organization creates and defines problems and then actively develops new knowledge to solve them" (Nonaka 1994, p. 14). How did the company actively develop new knowledge? The Director of Social Consciousness explains: "the company has somehow managed to get this far being very creative all these years and somewhat innovative, without a formal way…without any kind of formal innovation process or system". Though EF abides by no formal innovation procedures, its innovation comes from its company culture. Employees at EF claimed they felt included and empowered to participate in the company's progress: they felt invited to share their ideas, make suggestions and think about challenges they experience during their work and outside the company. Quinn explains that one of the main barriers to innovation is the bureaucratic barrier when top managers operate in isolation with little contact to employees and customers. By contrast, innovation occurs "When top executives appreciate innovation and manage their company's value system and atmosphere to support it" (Quinn 1987, p. 77). In one instance, the manager of the reuse programme remembered: "There was a woman … from the store. She was obsessed about what we were going to do with these damaged goods, so she would do a lot of research, and she found some machine that would felt fabric". According to Quinn, experts and fanatics are often pioneers in problem-solving (Quinn 1987). Innovation in this case came from an employee rather than a manager, a bottom-up approach to innovation, in which innovation is suffused in the foundations of the company culture. According to Westley, employee enthusiasm is dependent on their capacity to communicate with their managers (Westley 1990). The better the communication, the more energetic and empowered the employee. More importantly, "The ability of any organization to be cohesive depends on the structure and quality of its communication system" (Westley 1990, p. 337). Communication creates a positive, optimistic energy in the company itself, and it provides employees the possibility to innovate.

12 An Analysis of the Operation of the Take-Back Programme

The recycling factory was visited and analysed according to a variation on a multi-moment observation method to clarify EF's take-back and recycling programme at the operational level. The following grid (Table 1) describes each working process of the take-back and recycling process, as well as risks and actions taken. Analysing the operational level of the EF take-back programme details the many steps involved in a take-back programme. The aims are to identify key issues and factors which need to be addressed at multiple levels to improve the existing programme, and to determine how manufacturers can lay the foundations for adopting their own take-back programmes.

13 Next Steps

The results of the multi-moment analysis show that although EF has established a successful and functional take-back programme, many issues remain to be addressed at an operational level. However, most of EF's challenges could be improved with a few main actions presented in this next section. These steps could be taken proactively by any company seeking to integrate circular economy principles into its business.

13.1 Barcode Technology

The most time-intensive aspect of the process is the time it takes to manually sort the returned garments. Barcode technology containing metadata optimized for sorting, such as garment size, shape, colour, fabric and production year would optimize the processing logistics, allowing inventory to be collected, sorted, managed and transported. This embedded data offers opportunities for innovative companies to develop sophisticated barcode technology, which could either be proprietary for each company, or industry-wide, better facilitating any future take-back programmes.

Table 1 Operational steps of the EF take-back programme

Job	Challenge	Action	Effect	Risk/Considerations
Department: retail store				
Take back clothes from customers	Will customers bring used clothes back?	Customer receives voucher when returning a garment; $5 per garment regardless of condition	Consumers receive non-expiring gift cards, this might lead to accumulating of gift cards	Accruing debt with customer credit
	Where to put/store the returned clothes?	Space in the store is limited, but each EF store has a "systems wall", with some space behind the wall to store material	Space is further limited by the volume of other material, but regular pick-up has so far avoided major problems	There might be a risk of garment donations in fall and Easter
	Will the return of used clothing require too much time from sales associates?	So far this problem has not occurred. Usually, customers bring products but also want to look in the store for something new	No effect	Long processing wait times might deter customers from returning garments or from being assisted for new sales
	How to ensure used garments do not get in contact with new garments? (risking contamination)	When a customer brings used clothes they will be put in a plastic bag, if they are not already in one. Employees seal the bag	Additional environmental burden from countless plastic bags	Might consider reusable bags right from the beginning
	How to organize the clothing return to the warehouse?	Once the store has compiled a few bins, clothes in their plastic bags will be returned to the warehouse when new shipments arrive	Additional transportation and handling costs; each store defines when it has enough material to send	Additional costs and CO_2 emissions; stores might have different perceptions when they have enough material for return
	How will retail partners participate in the take-back programme?	External partners are excluded, garments are only taken back in EF stores	The programme is limited	Customers might not be aware that the take-back programme is only offered in EF store, might seek out a retail partner and be rebuffed
	What will be done if customers return products not from EF?	If this is recognized in the store, those garments will be refused	Additional sorting criteria	Customer might be offended

(continued)

Table 1 (continued)

Department: circular economy team operational process facility—mending area, recycling centre and warehouse in Irvington

Garments arrive in their plastic bags and are brought into the facility in bins	How many garments will arrive? Will the number be about the same all year?	The facility began sorting operation in 2010, recycling operation in July 2017. There is limited data available for the number of garments arriving per year, per month, per week, and whether the numbers will remain similar in the next years. EF is developing better tracking of weekly incoming and outgoing	There might be seasonal peaks in garment donations	Seasonal peaks may require flexible space and labour force. Risk is recognized by EF in their efforts to achieve more accurate tracking
	Can all the garments which arrive be processed/sorted?	About 2000 garments can be processed per week if more clothes arrive in-between storage required, if not enough material donated the labour will be short of work if there is no additional overflow material to process	Garments might pile up and occupy additional space. Without sufficient inventory on stock, employees might have no work	Even if insufficient material is returned, the facility still needs to be operated, which produces costs. If too much material cannot be processed, additional space is required which also produces additional costs
Bags are opened, garments are taken out and sorted by material, by season, by saleability and by brand (non-EF garments will be donated)	What is the condition of the returned garments?	The more garments which are not good enough for reuse, the more material needs recycling	Handling capacity in the recycling section might vary. Garments might remain in the facility over a long period	Garments of differing quality lead to variable income and increase recycling costs
	How can marks and defects be determined in the fastest way with minimal oversight?	Each garment is inspected by a trained employee and its condition determined. Very labour intensive. Defects are not marked since garment will be washed	The process is time-consuming and labour intensive	High costs. Might lead to back-up of returned products, requiring additional sorting
Off-season garments will not be washed but stored as inventory in colour-coded bags, sorted by product groups	Will there be enough storage space for off-season garments?	The interim storage requires an additional process	Additional costs for space and labour but it is unclear how much value the garments have	Since the material is not washed, insects could damage the garments and reduce the value

(continued)

Table 1 (continued)

In-season garments are laundered in-house or from contractor to clean them and remove stains	Which stains will be removed with the washing process?	Currently, no policy in place regarding stains; however, special stain treatments might achieve better results and increase product value. Currently, no data documents how much stains will disappear after laundry	It is questionable if stains can be removed after laundry since the EF customer takes care of her clothes and has probably washed and dry-cleaned them before	More information and data is required about the stains
	How much time and handling is necessary for laundry process? What are the costs?	A fixed laundry process is critical in a smooth operation process. Due to the operation time of each stage of the washing programme, if garments require additional treatment they will incur additional handling costs	The process flow is not guaranteed, and the volume of garments requiring laundering will vary	Time and capacity analysis needs to be done to organize the laundry process
Some in-season garments will be dry-cleaned if the material or garment is to be resold. Dry cleaning is typically more expensive than laundry and has higher environmental costs. If the garment requires recycling it will be washed	Is the dry clean really necessary?	Sorting garments for resale to be dry-cleaned	Garments might not be resold	Criteria for garments requiring dry cleaning must be set, especially if there is no longer a care label affixed to the garment
Returned garments will require another sorting process. The condition of every garment will need to be determined	How to define the level of use for a garment?	Workers must be trained to understand the required level of perfection	The level of perfection requires standards	Guidelines and standards have to be developed
	Will the inspection accurately determine whether a garment is in perfect condition or if it is a "second"? How is damage determined?	Since there is only one inspection, there is nothing else which can be done, except adding another inspection before delivery	Adding another inspection requires time and increases costs	Since another inspection might be too costly, stores require information what to do in case they receive a damaged garment
	Does the garment still have its labels?	Labels might need to be replaced, but require product knowledge	Risk of adding incorrect labels	Might lead to false care from consumer, might lead to disappointment

(continued)

Table 1 (continued)

Garments with little defects will be mended	How can the mending be done so that the garment becomes a sellable product?	Repairing the garment with the best possible mending technique	Some garments will turn out well and sell, others are mended but not sellable	Important to connect with sales and find out what kind of mending produces sellable garments. For example, should the mending thread be dyed to match or in contrast? Develop standards
Damaged garments (with little/big holes) will be processed in the maker space	How obvious is the damage? How can the defect be made visible?	If the damage is not obvious, the next sorting process must search again for the damage	Little defects might require additional time in the sorting process	Spending too much time in sorting, consider marking the defects
Garments in perfect condition and "seconds" will be separated and will be stored in the "warehouse section"	How long will these garments be stored in the warehouse before they can be delivered to a store?	Stores request used clothing or pick it up themselves depending on their proximity to storage facility	The longer garments are stored, the greater the costs. Stores might be sent material that the clientele of that particular store do not seek	Additional storage time increases the risk of damage to garments. For example, some garments might lose shape if hanging too long
	What is on stock in the warehouse? How to know what's available?	Currently, products are organized by product groups, size, and colour. The system is visually sorted by humans without an online system using product numbers	Since the warehouse is organized, an employee with retail experience can select pieces manually	Deliverable clothing volumes remain to be calculated. The wrong material might be sent to stores that does not sell, requiring the product to be reprocessed
Deliveries for stores with perfect garments will be put together for shipment	What kind of products should be chosen for resale?	Garments are put together and are waiting for next delivery to store—referred to as "picking"	Since the warehouse is organized, an employee with retail experience can select pieces manually Includes the risk of picking the wrong garments for sales. Who is responsible for putting the deliveries together?	What are the criteria: For example, will the garments be picked by season? Is it better to have the same style in a few sizes? Relying on skilled pickers might create a bottleneck in distribution
"Seconds" are sold during warehouse sales a few times per year, currently only in Seattle location	How long will the seconds be stored before they can be sold? How much space will they require?	Organizing and Logistic of bringing the "seconds" to Seattle	Additional time for handling and costs for transport	Although seconds sell well, there might be not enough revenue due to the costs of handling, transport and storage

(continued)

Table 1 (continued)

Department: circular economy team operational process facility—sorting area, recycling centre and warehouse in Irvington

Garments with stains and holes arrive at the sorting space and are organized by the fabric content (or fabric groups, because in some categories there are very few material), style, colour and occasionally size. The material groups are put together in batches, come in bags and are stored	Will there be enough similar material with matching colours?	Insufficient matching material requires additional searching for new material, further complicating the process	Different piles require additional space and time for sorting. Additional costs for plastic bags for pre-sorted materials	The various piles might become disorganized
	At which place of the garment is the whole or stain can the standard pattern be applied?	Finding the right pattern or searching for additional material	Search for a pattern and additional material requires additional time	Without a workflow, the process cannot scale into an industrial production
If designated for Resewn, the garment is deconstructed (seams are opened, each single garment piece is cut and sewn together)	How can the seams be opened, the pieces cut and sewn together in the fastest way while still producing an attractive garment?	Currently, there is not enough labour, so the work is well done but not in an industrial way	Little output and high costs	Without a workflow, the process cannot scale into an industrial production
Labelling and storage of Resewn garments	For which season are those garments made? How do they blend in the store with the new deliveries?	75,000 pieces are held in clear plastic bags on stock	Although EF has a large stockpile, they currently produce a small volume of Resewn garments, limiting their distribution to every store. Stores might not get enough material, limiting visibility for the Resewn concept	The produced garments might be too expensive and not financially viable. If not enough garments are delivered, sales are limited

Department: circular economy team operational process facility—felting area, recycling centre in Irvington

Deconstructed garment pieces are felted together, tests with fabric scraps are being conducted	How can the pieces be felted together in the fastest and most beautiful way?	A trained designer puts the layers together and prepares the textile for felting. The process takes a long time	If a highly trained person is doing this, job costs remain high, and the existing two machines won't be used to full capacity	Not enough output; costly machinery. Consider training of workers with lower education

13.2 Defining Reusability Standards and Criteria

While mending is generally preferable to recycling, because the garment is kept longer at a higher level of use, this process is only practical when the garment can be sold for reuse after the mending. Therefore, criteria must be developed to guide whether and how a garment should be mended. Clear guidelines for the sorting criteria of the received materials are also needed.

13.3 Making the Recycling Programme Financially Viable

Scaling the recycling initiative increases costs. While the reselling of the used garments is a great source for revenue and is promising due to an overall growth rate of the second-hand market, the main challenge in the future might be to collect more used garments good enough for reuse. A helpful strategy could be to convince partners to expand the EF take-back programme to retail partners and specialty stores. Further, this emphasizes the need to have the take-back programme financially independent from the reuse programme. Therefore, recycling operations must be optimized and scaled appropriately for the company to at least break-even.

13.4 Measuring the Environmental Success

Since EF is driven to realize a circular economy approach and to reduce its environmental footprint, it is important to assess the environmental impact of its take-back programme. A circular business model is based on reducing, reusing and recycling, and each of these goals, according to Bocken et al. (2016) can be examined according to their slowing, narrowing and closing effects on the environmental impact. The main effect of EF's take-back programme is to extend the life cycle of the garments—which slows consumption and resource need, but it is unclear to what degree this reduces consumption. The narrowing effect, which is based on manufacturing efficiencies, can be applied in various ways: for example, in the production process of the new garment. The line has an environmental impact because it requires energy and is, therefore, an increase in environmental burden. The current process for each garment involves washing and cleaning upon arrival at the recycling facility, but often consumers have washed their garments before donations. Skipping this process might provide opportunities to reduce environmental impact. To determine the effect of the washing and dry clean operation, including the transportation between the various facilities, an environmental assessment is needed.

The felting operation comes closest to closing the loop because the process produces a new fabric. Closing the loop would entail the making of a new fibre from the recycled garments (fibre-to-fibre recycling). However, felting also requires energy and new machinery. It prevents material going into the waste stream which is a clear environmental benefit, but it also entails environmental costs. The Resewn line represents a new manufacturing process, the environmental effects of which remain to be assessed.

14 Conclusion

EF is an industry leader in its circular economy approach to luxury fashion apparel. While it faces daily challenges at the operational level, EF's approach to a circular economy business model has changed the entire EF enterprise. Fundamental changes in one area of the business led to a profound restructuring in operations of the whole company. Amit and Zott claim that new business models often affect the entire enterprise (Amit and Zott 2001). EF's circular economy approach led the company to innovate in all five types of innovation described by Schumpeter (1982): first, EF developed new products such as pillows and purses made of felted materials; second, EF developed new production methods such as the overdying of silk garments with natural dyes; third, the sourcing team sought new suppliers for reclaimed fibres and eventually exchanged fibres like Rayon with Tencel, which can be produced in a closed loop; fourth, EF reorganized the business to incorporate a take-back programme in the development of its circular store, the Lab and forthcoming Brooklyn store. Finally, EF gained new customers and access to new markets by offering EF clothing at discounted (though nonetheless profitable) prices. Going forward, the company should seek to expand the programme to all partners in the supply chain and retail channels.

A circular economy approach aims to increase economic growth without environmental depredation. For EF, a circular economy approach has increased business through unique products aimed at loyal and new customers, requiring a limited amount of additional resources. The EF Renew programme has expanded the product line of EF with affordable luxury garments for the mass market and with new felted products for the luxury consumer. However, the full environmental benefit remains unclear. While the company collects about 170,000 garments per year, this represents only 2% of the total garments sold each year by the company. Even EF's business model remains linear until scaled. Moreover, collecting and diverting the material from the landfill is not a net saving for the environment, since the new operations require energy and produce waste. Therefore, further environmental assessment is necessary. The biggest opportunity in this new business model to extend the business without increasing resource scarcity and environmental degradation has been achieved, but further research is necessary to assess and evaluate the environmental benefit. Some strategic decisions could also help to make the programme more efficient and more financially viable, such as new barcode technology, the development of standards and criteria to evaluate the received material, better production planning and manufacturing processes.

A sustainable circular fashion economy requires socially conscious companies led by strong entrepreneurship invested in innovation at every level of its company structure. To develop a circular fashion economy, companies must strive to integrate several aspects under a sustainable approach: the business model, the product, the design, the manufacturing, the customer base and the retail operations.

Appendix 1: Cultivating a Circular Economy Mindset in a Retail Space

The Lab Store: This store is unique among most fashion retail stores in that it seeks to explain a circular economy fashion model to customers and still offers the effect of a luxury store. The sales associate explains: "She [the customer] really gets to understand the whole story, it represents all the different aspects of the life of her clothes". At the store entrance, for instance, visitors are greeted by a display of reclaimed cashmere sweaters set next to a garment from the sample collection (pieces which have been developed but didn't get into production) and finally pieces from the current product line. The Lab also offers a maker space with a working table and sewing machine where classes for knitting and sewing are held and customers have the opportunity to purchase manufacturing ends of yarn and fabrics as well as one-of-a-kind remade pieces. There is a sales section in one corner and a "Renew" section in another, where all used and seasonably appropriate garments in perfect condition are sold.

Appendix 2: Interview Participants and Questions

Director of Social Consciousness

- Could you please introduce yourself? What is your name, your position, and how many years have you worked at the company?
- How does EF innovate?
- Can you describe EF's actions related to fibre recycling?
- What is the difference between the circular economy team and the sustainability team? How is the circular economy team embedded in the organization?
- Can you describe how the circular economy approach started and how it has evolved?
- What are the challenges to make the circular economy model financially viable? What other challenges need to be overcome?
- Is the reuse programme a viable business component of the EF company?
- What are the opportunities for Eileen Fisher resulting from its circular economy work? What other business expansions has EF considered?
- What has EF's B-Corp certification meant for the enterprise of EF?

- Can you explain Eileen Fisher's idea behind the Employee Staff Ownership Plan?
- What is the biggest challenges for EF in terms of sustainability? Which role does the supply chain play in terms of sustainability?

Manager of the Take-back programme

- Could you please introduce yourself? What is your name, your position, and how many years have you worked at the company?
- Can you take me through the development of EF's take-back programme? What were some of the main drivers for the programme?
- What have been some of the notable developments in the take-back programme?
- How many jobs did the programme create?
- Could you describe EF's current business model in relation to its remake programme?
- Does the remake programme represent a new business model for EF? What about the reuse programme?
- Do you think other brands will offer remake and reuse programmes in the future?
- Do you think these programmes will become essential components of future business models?
- Do you think retail stores will sell second-hand clothing alongside their new clothes in-store?
- How does EF plan to scale its remake programme?

Sales Associates in Irvington and in New York

- Could you please introduce yourself? What is your name, your position, and how many years have you worked at the company?
- Can you talk me through a customer's typical first experience navigating EF's Irvington store?
- How has EF sought to showcase its circular economy approach to customers in-store?
- What is the rationale behind EF's in-store signage?

Recycle Operation Manager

- Could you please introduce yourself? What is your name, your position, and how many years have you worked at the company?
- When did EF start the factory?
- Can you explain EF's system processes related to sorting, mending, cleaning and remaking of the returned EF garments?

Designer Reuse Programme

- Could you please introduce yourself? What is your name, your position, and how many years have you worked at the company?
- Can you explain EF's system processes related to sorting, mending, cleaning and remaking of the returned EF garments?

Wholesale Marketing Manager

- Could you please introduce yourself? What is your name, your position, and how many years have you worked at the company?
- What distinguishes EF garments as luxury products?
- How has EF sought to cultivate an international presence as a luxury brand?
- Could you give me an overview of EF's collections? What are the volumes?
- Can you describe the rationale behind EF's in-store sales events?
- In what ways does EF seek to promote the circular economy approach with its retail partners?
- How has EF sought to showcase its circular economy approach to customers in wholesale retail environment?
- What challenges must EF overcome to bring its circular economy approach to retail partners?
- What does EF see as the advantages of bringing this circular economy approach to the wholesale environment?
- Do the reuse and resown programmes reach new customer bases? Do the same customers who purchase EF garments also purchase the reuse line?
- What are some of the obstacles blocking wholesale partners from participating in the programme?
- How does EF distinguish between its own brand and luxury fashion brands?
- What role does sustainability play in its brand messaging?

Buyer for EF Retail Stores

- Could you please introduce yourself? What is your name, your position, and how many years have you worked at the company?
- Can you talk me through how the take-back programme works?
- Does the reuse, resown programme reach new customer bases? Do the same customers who purchase EF garments also purchase the reuse line?
- What are the price differences between the multiple EF lines?
- To what extent does EF consider the circular economy approach a part of its brand identity?
- To what extent do you think the EF customer is aware of the company's circular economy approach?
- What have been notable factors in the success of the EF brand?
- Can you describe the logistics of the EF renew line?
- Does the remake programme represent a new business model for EF? What about the reuse programme?
- How similar are the aesthetics between the original and the renewed EF garments?
- How has EF approached these differences in regards to its brand identity?
- Who are the ambassadors for EF?
- How does EF work to attract younger customers?
- How has EF worked to extend its product line?

- What distinguishes EF garments as luxury products?

Knitwear Designer

- Could you please introduce yourself? What is your name, your position, and how many years have you worked at the company?
- What is your background in designing sustainable fashion?
- What distinguishes EF garments as luxury products?
- How responsive is EF to fashion trends?
- What terms would you use to describe the EF clothing style?
- To what extent does a circular economy approach influence your design choices? Do you design with circularity in mind?
- Can you describe a typical working day for you? What are the departments you liaise with to develop the collection?

Sustainability Leader

- Could you please introduce yourself? What is your name, your position, and how many years have you worked at the company?
- What sort of environmental assessments have been conducted regarding EF's take-back programme?
- Can you elaborate on any of these assessments?
- What do you think is the biggest environmental impact of the EF take-back programme?

Facilitating Manager for Fabric R&D

- Could you please introduce yourself? What is your name, your position, and how many years have you worked at the company?
- What reclaimed fibres have been developed at EF and where are the problems with reclaimed fibres?
- What is the difference between developing a new fabric quality compared to a renewed fabric quality?
- How many fabric qualities based on reclaimed fibres is EF currently using?
- When did EF begin using reclaimed fabrics?
- Is there enough reclaimed fabric on the market for the luxury sector to scale sustainable textiles?

References

American Chemical Society. (2017). *Upcycling 'fast fashion' to reduce waste and pollution* [Press release]. Retrieved from https://www.eurekalert.org/pub_releases/2017-04/acs-uf030717.php.

Amit, R., & Zott, C. (2001). Value creation in e-business. *Strategic Management Journal, 22*(6–7), 493–520.

B-Corporations. *Declaration of independence.* Retrieved from https://www.bcorporation.net/what-are-b-corps/the-b-corp-declaration.

Bhardwaj, V., & Fairhurst, A. (2010). Fast fashion: Response to changes in the fashion industry. *The International Review of Retail, Distribution and Consumer Research, 20*(1), 165–173.

Birtwistle, G., & Moore, C. (2007). Fashion clothing—Where does it all end up? *International Journal of Retail & Distribution Management, 35*(3), 210–216.

Black, S. (2012). *The sustainable fashion handbook*. Thames and Hudson.

Bocken, N., Miller, K., & Evans, S. (2016). *Assessing the environmental impact of new Circular business models*. Paper presented at the Conference proceedings "New Business Models"—Exploring a changing view on organizing value creation. Toulouse, France.

Bye, E., & McKinney, E. (2007). Sizing up the wardrobe—Why we keep clothes that do not fit. *Fashion Theory: The Journal of Dress, Body & Culture, 11*(4), 483–498.

Conti, S. (2016). Burberry's bold move: To make shows direct to consumer. *WWD*. Retrieved from http://wwd.com/fashion-news/designer-luxury/burberry-runway-delivery-schedule-direct-consumer-10340340/.

Council for Textile Recycling. (2014). Retrieved from http://www.weardonaterecycle.org/about/issue.html.

Curwen, L. G., Park, J., & Sarkar, A. K. (2013). Challenges and solutions of sustainable apparel product development: A case study of Eileen Fisher. *Clothing and Textiles Research Journal, 31*(1), 32–47.

Ellen Mccarthur Foundation. (2017). *A new textiles economy: Redesigning fashion's future*. United Kingdom.

Fisher, E. (2017). *Our vision 2020*. Retrieved from https://www.eileenfisher.com/vision-2020/.

Fletcher, K. (2013). *Sustainable fashion and textiles: Design journeys*. Routledge.

Geng, Y., & Doberstein, B. (2008). Developing the circular economy in China: Challenges and opportunities for achieving 'leapfrog development'. *The International Journal of Sustainable Development & World Ecology, 15*(3), 231–239.

Ghisellini, P., Cialani, C., & Ulgiati, S. (2016). A review on circular economy: The expected transition to a balanced interplay of environmental and economic systems. *Journal of Cleaner Production, 114,* 11–32.

Hawley, J. (2009). Understanding and improving textile recycling: A systems perspective. *Sustainable textiles: Life cycle and environmental impact. Woodhead Publishing series in textiles* (Vol. 98, pp. 179–200).

Henninger, C. E., Ryding, D., Alevizou, P. J., & Goworek, H. (2017). *Introduction to sustainability in fashion sustainability in fashion* (pp. 1–10). Springer.

Jensen, J. (2012). *Waste audit services project final summary report*. Retrieved from Halifax: http://putwasteinitsplace.ca/uploads/file/rrfb/RRFB-Waste_Audits_Summary-043012-web.pdf.

Kant Hvass, K. (2014). Post-retail responsibility of garments—a fashion industry perspective. *Journal of Fashion Marketing and Management, 18*(4), 413–430.

Karaosman, H. (2016). *Sustainability in fashion*.

Koh, H. (2017). A way to repeatedly recycle polyester has just been discovered. Retrieved from http://www.ecobusiness.com/news/a-way-to-repeatedly-recycle-polyester-has-just-been-discovered/.

Laitala, K. (2014). Consumers' clothing disposal behaviour—a synthesis of research results. *International Journal of Consumer Studies, 38*(5), 444–457.

Lang, C., Armstrong, C. M., & Brannon, L. A. (2013). Drivers of clothing disposal in the US: An exploration of the role of personal attributes and behaviours in frequent disposal. *International Journal of Consumer Studies, 37*(6), 706–714.

Lieder, M., & Rashid, A. (2016). Towards circular economy implementation: A comprehensive review in context of manufacturing industry. *Journal of Cleaner Production 115*, 36–51.

MacArthur, E. (2013). *Towards the circular economy, economic and business rationale for an accelerated transition*. Ellen MacArthur Foundation: Cowes, UK.

Maybelle, M. (2015). Throwaway fashion: Women have adopted a 'wear it once culture', binning clothes after only a few wears (so they aren't pictured in same outfit twice on social media). *Mailonline*. Retrieved from http://www.dailymail.co.uk/femail/article-3116962/Throwaway-fashion-Women-adopted-wear-culture-binning-clothes-wears-aren-t-pictured-outfit-twice-social-media.html.

McCourt, A., & Perkins, L. (2015). Designing for the circular economy: Cradle to cradle design. In J. Hethorn & C. Ulasewicz (Eds.), *Sustainable fashion: What's next? A conversation about issues, practices and possibilities*. Bloomsbury Publishing, USA.

Moon, K. K.-L., Lai, C. S.-Y., Lam, E. Y.-N., & Chang, J. M. (2015). Popularization of sustainable fashion: barriers and solutions. *The Journal of the Textile Institute, 106*(9), 939–952.

Nonaka, I. (1994). A dynamic theory of organizational knowledge creation. *Organization Science, 5*(1), 14–37.

Okonkwo, U. (2016). *Luxury fashion branding: Trends, tactics, techniques*. Springer.

Pedersen, E. R. G., & Andersen, K. R. (2015). Sustainability innovators and anchor draggers: A global expert study on sustainable fashion. *Journal of Fashion Marketing and Management, 19*(3), 315–327.

Pedersen, E. R. G., & Gwozdz, W. (2014). From resistance to opportunity-seeking: Strategic responses to institutional pressures for corporate social responsibility in the Nordic fashion industry. *Journal of Business Ethics, 119*(2), 245–264.

Phau, I., & Prendergast, G. (2000). Consuming luxury brands: The relevance of the 'rarity principle'. *The Journal of Brand Management, 8*(2), 122–138.

Quinn, J. B. (1987). Managing innovation: Controlled chaos. *Harvard Business Review.* 研究 技術 計画, *2*(4), 485.

REFA-Methodenlehre. (1978). Teil 2: Datenermittlung. *Auflage (München 1973)*.

Schumpeter, J. A. (1982). *The theory of economic development: An inquiry into Profits, Capital, Credit, Interest, and the Business Cycle* (1912/1934, p. 244). Transaction Publishers, 1 January 1982.

SCS Global Services. (2014). *Responsible source—Textiles synthetic materials for textiles.* Retrieved from https://www.scsglobalservices.com/responsible-source-textiles.

Smith, J. (2012, April 17–19). *How can understanding the consumer make fashion more sustainable?* Paper presented at the 10th European academy of design conference–crafting the future. Gothenburg, Sweden.

Smith, P., Baille, J., & McHattie, L.-S. (2017). Sustainable design futures: An open design vision for the circular economy in fashion and textiles. *The Design Journal, 20*(sup1), S1938–S1947.

Stahel, W. R. (2016). Circular economy: A new relationship with our goods and materials would save resources and energy and create local jobs. *Nature, 531*(7595), 435–439.

Stall-Meadows, C., & Goudeau, C. (2012). An unexplored direction in solid waste reduction: Household textiles and clothing recycling. *Journal of Extension, 50*(5), 5RIB3.

Strähle, J., Shen, B., & Köksal, D. (2015). Sustainable fashion supply chain: Adaption of the SSI-index for profiling the sustainability of fashion companies. *The International Journal of Business & Management, 3*(5), 232.

Sull, D., & Turconi, S. (2008). Fast fashion lessons. *Business Strategy Review, 19*(2), 4–11.

The American Apparel & Footwear Association. (2012). *AAFA releases apparel stats 2012 report.* Retrieved from https://www.wewear.org/aafa-releases-apparelstats-2012-report/.

Textile Exchange. (2017). *Integrity and Standards.* Retrieved from http://textileexchange.org/integrity/.

Uitdenbogerd, D. E., Brouwer, N. E., & Groot-Marcus, J. P. (1998). *Domestic energy saving potentials for food and textiles: An empirical study.* The Netherlands: Retrieved from Wageningen.

United States Environmental Protection Agency. (2013). *Textiles.* Retrieved from http://www.epa.gov/epawaste/nonhaz/municipal/msw99.htm.

Vancity. (2016). *Thrift score: An examination B.C.'s second-hand economy.* Retrieved from http://www.marketwired.com/press-release/report-bcs-second-hand-economy-rings-up-1-billion-in-sales-annually-2172487.htm.

Victor Group Inc. (2008). *Victor Innovator*. Retrieved from http://www.victor-innovatex.com.

Watson, D., Plan, M., Eder-Hansen, J., Tärneberg, S. (2017). *A call to action for a circular fashion system*. Retrieved from Copenhagen https://www.copenhagenfashionsummit.com/wp-content/uploads/2017/04/GFA17_Call-to-action_Poluc-brief_FINAL_9May.pdf.

Weber, S. (2015). *How consumers manage textile waste*. Retrieved from Waterloo.

Westley, F. R. (1990). Middle managers and strategy: Microdynamics of inclusion. *Strategic Management Journal, 11*(5), 337–351.

Yin, R. K. (2013). *Case study research: Design and methods*. Sage publications.

Zamani, B., Svanström, M., Peters, G., & Rydberg, T. (2015). A carbon footprint of textile recycling: A case study in Sweden. *Journal of Industrial Ecology, 19*(4), 676–687.

Sabine Weber, MES, B.A. (Hons.), is a Professor at Seneca College in the School of Fashion, Toronto, Canada. Her research topics are sustainable fashion, textile waste and social innovation. Her main interest is how social innovation can help to make the fashion industry a circular system. Sabine Weber has studied Apparel Engineering in Germany and worked for many years in the fashion industry as international buyer and team leader for different brands. In 2015 she graduated at the University of Waterloo in the School of Environment and Resource Sustainability (SERS) on how consumers manage textile waste. She is currently working on her PhD in Social Innovation and is part of the leading Committee of Ontario's Textile Diversion Coalition.

Printed in the United States
By Bookmasters